CLONE

The Baby Doctors
Probing the Limits of Fetal Medicine

Sex in America: A Definitive Survey,
with Robert T. Michael, John H. Gagnon, and
Edward O. Laumann

Combating the Number One Killer:
The Science Report on Heart Research,
with Jean L. Marx

The High Blood Pressure Book:
A Guide for Patients and Their Families,
with Edward D. Fries, M.D.

CLONE

THE ROAD TO DOLLY, AND THE PATH AHEAD

GINA KOLATA

WILLIAM MORROW AND COMPANY, INC.

NEW YORK

Library of Congress Cataloging-in-Publication Data

Kolata, Gina Bari. 1948–
    Clone : the road to Dolly, and the path ahead / Gina Kolata.
        p.   cm.
    Includes bibliographical references and index.
    ISBN 0-688-15692-4
    I. Cloning—Moral and ethical aspects. I. Title.
QH442.2.K65   1998
174'.957—dc21
                                                                        97-29132
                                                                        CIP

Printed in the United States of America

First Edition

1   2   3   4   5   6   7   8   9   10

BOOK DESIGN BY JO ANNE METSCH

www.williammorrow.com

To Bill, Therese, and Stefan

ACKNOWLEDGMENTS

In the course of writing this book, I have met extraordinary people and had generous help from scientists and others that far exceeded my expectations.

Every living scientist who is quoted deserves my thanks for graciously agreeing to interviews and helping me piece together a history that has never been recorded in books. I would like to thank, in particular, Robert McKinnell, Randall Prather, Steen Willadsen, and Shirley Tilghman for checking sections of the book for scientific accuracy. I also thank Davor Solter for providing me with his personal files of invaluable historical material and Lee Silver for patiently answering innumerable technical questions.

I could not have written this book without the research help I received from Alexandra Nyberg, who enthusiastically and assiduously found even the most obscure references.

And, as always, my husband, Bill, provided invaluable assistance by reading every word of my manuscript and giving me his insightful comments.

CONTENTS

CLONE

1

A CLONE IS BORN

Many people wonder if this is a miracle for which we can thank God, or an ominous new way to play God ourselves.

—NANCY DUFF

Princeton Theological Seminary

On a soft summer night, July 5, 1996, at 5:00 P.M., the most famous lamb in history entered the world, head and forelegs first. She was born in a shed, just down the road from the Roslin Institute in Roslin, Scotland, where she was created. And yet her creator, Ian Wilmut, a quiet, balding fifty-two-year-old embryologist, does not remember where he was when he heard that the lamb, named Dolly, was born. He does not even recall getting a telephone call from John Bracken, a scientist who had monitored the pregnancy of the sheep that gave birth to Dolly, saying that Dolly was alive and healthy and weighed 6.6 kilograms, or 14.5 pounds.

It was a moment of remarkable insouciance. No one broke open champagne. No one took pictures. Only a few staff members from the institute and a local veterinarian who attended the birth were present. Yet Dolly, a fluffy creature with grayish-white fleece and a snow-white face, who looked for all the world like hundreds of other lambs that dot the rolling hills of Scotland, was soon to change the world.

When the time comes to write the history of our age, this quiet birth, the creation of this little lamb, will stand out. The events that change history are few and unpredictable. In the twentieth century, there was the discovery of quantum theory, the revolutionary finding by physicists that the normal rules of the visible world do not apply in the realm of the atom. There was Einstein's theory of general relativity, saying that space and time can be warped. There was the splitting of the atom, with its promise of good and evil. There was the often-overlooked theorem of mathematician Kurt Gödel, which said that there are truths that are unknowable, theorems that can be neither proved nor disproved. There was the development of computers that transformed Western society.

In biology and medicine, there was the discovery of penicillin in the 1940s, and there was James Watson and Francis Crick's announcement, in 1953, that they had found the structure of DNA, the genetic blueprint. There was the conquest of smallpox that wiped the ancient scourge from the face of the earth, and the discovery of a vaccine that could prevent the tragedy of polio. In the 1980s, there was the onslaught of AIDS, which taught us that plagues can afflict us still.

In politics, there were the world wars, the rise and fall of communism, and the Great Depression. There is the economic rise of Asia in the latter part of the century, and the ever-shifting balance of the world's powers.

But events that alter our very notion of what it means to be human are few and scattered over the centuries. The birth of Dolly is one of them. "Analogies to Copernicus, to Darwin, to Freud, are appropriate," said Alan Weisbard, a professor of law and medical ethics at the University of Wisconsin. The world is a different place now that she is born.

Dolly is a clone. She was created not out of the union of a sperm and an egg but out of the genetic material from an udder cell of a six-year-old sheep. Wilmut fused the udder cell with an egg from another sheep, after first removing all genetic material from the egg. The udder cell's genes took up residence in the egg and directed it to grow and develop. The result was Dolly, the identical twin of the original sheep that provided the udder cells, but an identical twin born six years later. In a moment of frivolity, as a wry joke, Wilmut named her Dolly after Dolly Parton, who also was known, he said, for her mammaries.

Until Dolly entered the world, cloning was the stuff of science fiction. It had been raised as a possibility decades ago, then dismissed, relegated to the realm of the kooky, the fringy, something that serious scientists thought was simply not going to happen anytime soon.

Yet when it happened, even though it involved but one sheep, it was truly fantastic, and at the same time horrifying in a way that is hard to define. In 1972, when Willard Gaylin, a psychiatrist and the founder of the Hastings Center, an ethics think tank, mistakenly thought that science was on the verge of cloning, he described its awesome power: "One could imagine taking a single sloughed cell from the skin of a person's hand, or even from the hand of a mummy (since cells are neither 'alive' nor 'dead,' but merely intact or not intact), and seeing it perpetuate itself into a sheet of skin tissue. But could one really visualize the cell forming a finger, let alone a hand, let alone an embryo, let alone another Amenhotep?"

And what if more than one clone is made? Is it even within the realm of the imaginable to think that someday, perhaps decades from now, but someday, you could clone yourself and make tens, dozens, hundreds of genetically identical twins? Is it really science fiction to think that your cells could be improved before-hand, genetically engineered to add some genes and snip out others? These ideas, that so destroy the notion of the self, that touch on the idea of the soul, of human identity, seemed so implausible to most scientists that they had declared cloning off-limits for discussion.

Even ethicists, those professional worriers whose business it is to raise alarms about medicine and technology, were steered away from talk of cloning, though they tried to make it a serious topic. In fact, it was one of the first subjects mentioned when the bioethics field came into its own in the late 1960s and early 1970s. But scientists quashed the ethicists' ruminations, telling them to stop inventing such scary scenarios. The ethicists were informed that they were giving science a bad name to raise such specters as if they were real possibilities. The public would be frightened, research grants might dry up, scientists would be seen as Frankensteins, and legitimate studies that could benefit humankind could be threatened as part of an anti-science backlash.

Daniel Callahan, one of the founders of the bioethics movement and the founder, with Gaylin, of the Hastings Center, recalled that when he and others wanted to talk about cloning, scientists pooh-poohed them. They were told, he said, that "there was no real incentive for science to do this and it was just one of those scary things that ethicists and others were talking about that would do real harm to science."

Now, with the birth of Dolly, the ethicists were vindicated. Yes, it was a sheep that was cloned, not a human being. But there was nothing exceptional about sheep. Even Wilmut, who made it clear that he abhorred the very idea of cloning people, said that there was no longer any theoretical reason why humans could not clone themselves, using the same methods he had used to clone Dolly. "There is no reason in principle why you couldn't do it." But, he added, "all of us would find that offensive."

*   *   *

The utterly pragmatic approach of Wilmut and many other scientists, however, ignores the awesome nature of what was accomplished. Our era is said to be devoted to the self, with psychologists and philosophers battling over who can best probe the nature of our identities. But cloning pares the questions down to their essence, forcing us to think about what we mean by the self, whether we are our genes or, if not, what makes us *us*. "To thine own self be true" goes the popular line from Shakespeare—but what is the self?

We live in an age of the ethicist, a time when we argue about pragmatism and compromises in our quest to be morally right. But cloning forces us back to the most basic questions that have plagued humanity since the dawn of recorded time: What is good and what is evil? And how much potential for evil can we tolerate to obtain something that might be good? We live in a time when sin is becoming one of those quaint words that we might hear in church but that has little to do with our daily world. Cloning, however, with its possibilities for creating our own identical twins, brings us back to the ancient sins of vanity and pride: the sins of Narcissus, who so loved himself, and of Prometheus, who, in stealing fire, sought the powers of God. In a time when we hear rallying cries of reproductive freedom, of libertarianism and the rights of people to do what they want, so long as they hurt no one else, cloning, by raising the possibility that people could be made to order like commodities, places such ideas against the larger backdrop of human dignity.

So before we can ask why we are so fascinated by cloning, we have to examine our souls and ask, What exactly so bothers many of us about trying to replicate our genetic selves? Or, if we are not bothered, why aren't we?

We want children who resemble us. Even couples who use donor eggs because the woman's ovaries have failed or because her eggs are not easily fertilized, or who use donor sperm because the man's sperm is not viable, peruse catalogs of donors to find people who resemble themselves. We want to replicate ourselves. Several years ago, a poem by Linda Pastan, called "To a Daughter Leaving Home," was displayed on the walls of New York subways. It read:

> Knit two, purl two,
> I make of small boredoms
> a fabric
> to keep you warm.
> Is it my own image
> I love so
> in your face?
> I lean over your sleep,
> Narcissus over
> his clear pool,
> ready to fall in—
> to drown for you
> if necessary.

Yet if we so love ourselves, reflected in our children, why is it so terrifying to so many of us to think of seeing our exact genetic replicas born again, identical twins years younger than we? Is there a hidden fear that we would be forcing God to give us another soul, thereby bending God to our will, or, worse yet, that we would be creating soul-less beings that were merely genetic shells of humans? After all, in many religions, the soul is supposed to be present from the moment of conception, before a person is born and shaped by nurture as well as nature. If a clone is created, how could its soul be different from the soul of the person who is cloned? Is it possible, as molecular biologist Gunther Stendt once suggested, that "a human clone would not consist of real persons but merely of Cartesian automata in human shape"?

Or is it one thing for nature to form us through the vagaries of the genetic lottery, and another for us to take complete control, abandoning all thoughts of somehow, through the mixing of genes, having a child who is like us, but better? Normally, when a man and a woman have a child together, the child is an unpredictable mixture of the two. We recognize that, of course, in the hoary old joke in which a beautiful but dumb woman suggests to an ugly but brilliant man that the two have a child. Just think of how wonderful the baby would be, the woman says, with my looks and your brains. Aha, says the man. But what if the child inherited *my* looks and *your* brains?

Theologians speak of the special status of a child, born of an act of love between a man and a woman. Of course, we already routinely employ infertility treatments, like donor eggs, semen banks, and frozen embryos, that have weakened these ties between the parents and the child. But, said Gilbert Meilaender, a Lutheran theologian, cloning would be "a new and decisive turn on this road." Cloning entails the *production,* rather than the creation, of a child. It is "far less a surrender to the mystery of the genetic lottery," he said, and "far more an understanding of the child as a product of human will."

Elliott Dorff, a rabbi at the University of Judaism in Los Angeles, said much the same thing. "Each person involved has to get out of himself or herself in order to make and have a child." But if a person can be reproduced through cloning, that self-surrender is lost, and there is danger of self-idolization.

Cloning also poses a danger to our notion of mortality, Dorff said. The biblical psalm says, "Teach us to number our days so that we can obtain a heart of wisdom," he recalled. "The sense that there is a deadline, that there is an end to all this, forces us to make good use of our lives."

In this age of entertainment, when philosophical and theological questions are pushed aside as too difficult or too deep, cloning brings us face-to-face with our notion of what it means to be human and makes us confront both the privileges and limitations of life itself. It also forces us to question the powers of science. Is there, in fact, knowledge that we do not want? Are there paths we would rather not pursue?

The time is long past when we can speak of the purity of science, divorced from its consequences. If any needed reminding that the innocence of scientists was lost long ago, they need only recall the comments of J. Robert Oppenheimer, the genius who was a father of the atomic bomb and who was transformed in the process from a supremely confident man, ready to follow his scientific curiosity, to a humbled and stricken soul, wondering what science had wrought.

Before the bomb was made, Oppenheimer said, "When you see something that is technically sweet you go ahead and do it." After the bomb was dropped on Hiroshima and Nagasaki, in a chilling speech at the Massachusetts Institute of Technology in 1947, he said: "In some sort of crude sense which no vulgarity, no humor, no overstatement can quite extinguish, the physicists have known sin; and this is a knowledge which they cannot lose."

As with the atom bomb, cloning is complex, multilayered in its threats and its promises. It offers the possibility of real scientific advances that can improve our lives and save them. In medicine, scientists dream of using cloning to reprogram cells so we can make our own body parts for transplantation. Suppose, for example, you needed a bone-marrow transplant. Some deadly forms of leukemia can be cured completely if doctors destroy your own marrow and replace it with healthy marrow from someone else. But the marrow must be a close genetic match to your own. If not, it will lash out at you and kill you. Bone marrow is the source of the white blood cells of the immune system. If you have someone else's marrow, you'll make their white blood cells. And if those cells think you are different from them, they will attack.

Today, if you need marrow, you have to hope that a sister, brother, parent, or child happens to have bone-marrow cells that are genetically compatible with your own. If you have no relative

whose marrow matches yours, you can search in computer data-
bases of people who have volunteered to donate their marrow, but
your chances of finding someone who matches you are less than
one in twenty thousand—or one in a million if your genetic type
is especially rare.

But suppose, instead, that scientists could take one of your
cells—any cell—and merge it with a human egg. The egg would
start to divide, to develop, but it would not be permitted to
divide more than a few times. Instead, technicians would bathe
it in proteins that direct primitive cells, embryo cells, to become
marrow cells. What started out to be a clone of you could grow
into a batch of your marrow—the perfect match.

More difficult, but not inconceivable, would be to grow solid
organs, like kidneys or livers, in the same way.

Another possibility is to create animals whose organs are perfect
genetic matches for humans. If you needed a liver, a kidney, or
even a heart, you might be able to get one from a pig clone that
was designed so it had human proteins on the surface of its organs.
The reason transplant surgeons steer away from using animal organs in humans, even though there is a dire shortage of human
organs, is that animals are so genetically different from people. A
pig kidney transplanted into a human is just so foreign that the
person's immune system will attack it and destroy it. But cloning
offers a different approach. Scientist could take pig cells, for example, and add human genes to them in the laboratory, creating pig
cells that were coated with human proteins. Then they could make
cloned pigs from those cells. Each pig would have organs that
looked, to a human immune system, for all the world like a
human organ. These organs could be used for transplantation.

Cloning could also be used to make animals that are living
drug factories—exactly the experiment that Ian Wilmut's sponsor,
a Scottish company called PPL Therapeutics, Ltd., wants to con-
duct. Scientists could insert genes into laboratory cells that would
force the cells to make valuable drugs, like clotting factors for

hemophiliacs. Then they could clone animals from those cells and create animals that made the drugs in their milk. The only step remaining would be to milk the clones and extract the drugs.

Another possibility would be to clone prize dairy cows. The average cow produces about fifteen thousand pounds of milk annually, but world champion milk producers make as much as forty thousand pounds of milk a year. The problem for breeders is that there are, apparently, so many genes involved in creating one of these phenomenal cows that no one has learned how to breed them the old-fashioned way. But if you had a cow that produced forty thousand pounds of milk a year, you could clone her and make a herd.

Zoologists might clone animals that are on the verge of extinction, keeping them alive and propagating when they might otherwise have vanished from the earth.

The possibilities are limitless, scientists say, and so, some argue, we should stop focusing on our hypothetical fears and think about the benefits that cloning could bring.

Others say that cloning is far from business as usual, far from a technical advance, and that we should be wary of heading down such a brambly path.

But was the cloning of Dolly really such a ground-shifting event? After all, the feat came as a climax to years of ever more frightening, yet dazzling, technological feats, particularly in the field of assisted reproduction. Each step, dreaded by some, cursed by others, welcomed by many more, soon grew to be part of the medical landscape, hardly worthy of comment. And so, with this history as background, some asked why, and how, anyone thought cloning could be controlled—or why anyone would want to. Besides, some asked, why was cloning any different in principle from some of the more spectacular infertility treatments that are accepted with hardly a raised eyebrow?

The infertility revolution began in 1978, when Louise Brown

was born in England, the world's first test-tube baby. After more than a decade of futile efforts, scientists finally had learned to fertilize women's eggs outside their bodies, allowing the first stages of human life to begin in a petri dish in a laboratory. The feat raised alarms at the time. It was, said Moshe Tendler, a professor of medical ethics and chair of the biology department at Yeshiva University, "matchmaking at its most extreme, two reluctant gametes trying to be pushed together whether they liked it or not."

But in vitro fertilization flourished despite its rocky start, nourished by the plaintive cries of infertile couples so unjustly condemned to be barren, and justified by the miracle babies—children who were wanted so badly that their parents were willing to spend years in doctors' offices, take out loans for tens of thousands of dollars, and take their chances of finally, ultimately, failing and losing all hope of having a child who bore their genes. The doctors who ran the clinics soothed the public's fears. In vitro fertilization was not horrifying, they said. It was just a way to help infertile couples have babies.

The federal government quickly got out of the business of paying for any research that even peripherally contributed to the manipulation of human embryos, but in vitro fertilization clinics simply did research on their own, with money from the fees they charged women for infertility treatments, and so the field advanced, beyond the purview of university science, with its federal grants and accompanying strict rules and regulations.

"There are no hard-and-fast rules; there is no legislation," said Arthur Wisot, the executive director of the Center for Advanced Reproductive Care in Redondo Beach, California. "This whole area of medicine is totally unregulated. We don't answer to anyone but our peers."

Nearly every year, the fertility clinics would take another step. Recently, they began advertising something they called intercytoplasmic sperm injection, or I.C.S.I., in which they could get usable sperm even from men who seemed to make none, or whose sperm cells were misshapen or immotile and simply unable to fertilize an egg. The scientists would insert a needle into a man's testicle and remove immature sperm, which were little more than raw genes. They would inject these nascent sperm into an egg to create an embryo. Medical scientists later discovered that many of these men had such feeble sperm because the genes that controlled their sperm production were mutated. When the sperm, carrying the mutated gene, were used to make a baby boy, the boy would grow up with the same mutations and he, too, would need I.C.S.I. to have a baby. Some scientists worried that there might be other consequences of such a mutation.

But the infertility doctors and many infertile couples were unconcerned by the possibility that this technique might be less of an unqualified boon than it at first appeared. And the I.C.S.I. advertisements continued unabated.

Infertility doctors also learned to snip a cell from a microscopic embryo and analyze it for genetic defects, selecting only healthy embryos to implant in a woman's womb. They learned that there is no age barrier to pregnancy: Women who had passed the age of menopause could still carry a baby if they used eggs from a younger woman, fertilized in a laboratory. Even women in their early sixties have gotten pregnant, and while some doctors have said they do not want to participate in creating such pregnancies, others say that it is really up to the women whether they want to become mothers at such an advanced age.

Infertility clinics are even learning to do the ultimate prenatal testing: fishing fetal cells out of a pregnant woman's blood and analyzing them for genetic defects. It is, said Tendler, "the perfect child syndrome. We can now take 5 cc of a woman's blood when she is seven to nine weeks pregnant, do 191 genetic probes on that cell, and decide whether that baby is going to make it or not."

The latest development involves methods to sort sperm, separating those sperm with Y chromosomes, which would create boys, from those with X chromosomes, which would create girls. Soon parents can have the ultimate control over the sex of their babies.

At the same time, molecular biologists learned to snip genes out of cells and to sew others in, engineering cells to order. Infertility clinics expect, before long, to be able to add genes to human embryos—or delete genes that could cause disease or disability—creating a perfect child before even implanting an embryo into a woman's womb.

At first, the feats of reproductive scientists were the objects of controversy and shock. But we have become accustomed to their achievements. And it is hard to argue against the cries that couples have a right to reproductive freedom. Many have suffered for years, yearning for a child of their own. If they want to create babies, and are paying with their own money, who has the right to tell them no?

These days, when infertility doctors introduce a new method to the public, or when their techniques disrupt what we have thought of as the natural order, there is, at first, a ripple of surprise, or sometimes dismay, but then that reaction fades and all we remember is that there seemed to be reports of one more incredible technological trick.

Even newspapers are becoming blasé. One Sunday in April, about six weeks after the cloning of Dolly was announced, I was attending a meeting of a federal commission that was assessing cloning. I crept out of the meeting to call a national news editor at *The New York Times* and inform him of the meeting's progress. He said there was something else he wanted to ask me about. There was a story out of Florida, he said, about a woman who just gave birth to her own grandchild. Was that news, he asked me?

I assured him that it was not news. Several years ago, another woman had done the same thing, and we'd reported it on page

1. The woman's daughter had been born with ovaries but not a uterus, so the mother carried the baby for the daughter. That story had come and gone, no longer even worth a raised eyebrow.

So when Dolly was born, in this age of ever-more-disarming scientific advances, some worried that her birth might be greeted with a brief shiver, then forgotten, like the woman who gave birth to her own grandchild. Leon Kass, a biochemist turned philosopher, at the University of Chicago, warned that to react as though cloning were just another infertility treatment would be to miss the point of Dolly. He worried that we may be too jaded by previous triumphs of technological wizardry to take cloning as seriously as we should. He quoted Raskolnikov, the protagonist of Fyodor Dostoyevsky's *Crime and Punishment:* "Man gets used to everything—the beast."

It is true, of course, that the revolution in infertility treatments set the stage for people to think about cloning a human. Were it not for the proficiency of doctors in manipulating human eggs and sperm, it would not be feasible to even think of transferring the chromosomes of an adult cell into a human egg. But there is an intellectual chasm between methods that result in a baby with half its genes from the mother and half from the father and cloning, which would result in a baby whose genes are identical to those of an adult who was cloned.

Human cloning, Kass said, would be "something radically new, both in itself and in its easily foreseeable consequences. The stakes here are very high indeed." Until now "we have benefited mightily from the attitude, Let technology go where it will and we can fix any problems that might arise later." But, he said, "that paradigm is open to question." Now we are "threatened with really major changes in human life, even human nature." And even if an absolute prohibition on cloning cannot be made effective, "it would at least place the burden on the other side to show the necessity" of taking this awesome step.

What is at issue, Kass said, "is nothing less than whether

human procreation is going to remain human, whether children are going to be made rather than begotten, and whether it is a good thing, humanly speaking, to say yes to the road which leads, at best, to the dehumanized rationality of *Brave New World.*" And so "What we have here is not business as usual, to be fretted about for a while and then given our seal of approval, not least because it appears to be unusual." Instead, he said, "the future of humanity may hang in the balance."

The cloning debate, Kass said, is so much more than just an argument about one more step in assisted reproduction. "This is really one of those critical moments where one gets a chance to think about terribly important things. Not just genetics and what is the meaning of mother and father and kinship, but also the whole relationship between science and society and attitudes toward technology." Cloning, he said, "provides the occasion as well as the urgent necessity of deciding whether we shall be slaves of unregulated progress and ultimately its artifacts or whether we shall remain free human beings to guide our technique towards the enhancement of human dignity."

He quoted the theologian Paul Ramsey: "Raise the ethical questions with a serious and not a frivolous conscience. A man of frivolous conscience announces that there are ethical quandaries ahead that we must urgently consider before the future catches up with us. By this he often means that we need to devise a new ethics that will provide the rationalization for doing in the future what men are bound to do because of the new actions and interventions science will have made possible. In contrast, a man of serious conscience means to say in raising urgent ethical questions that there may be some things that men should never do. The good things that men do can be made complete only by the things they refuse to do."

Yet if there is one lesson of cloning it is that there is no uniformly accepted way to think about the ethical questions that it elicits, and no agreement, even among the most thoughtful and well-informed commentators, about what is right and what is wrong. Many—but by no means all—theologians tended to condemn the notion of human cloning. Many ethicists were similarly repelled, but others asked instead, who would be harmed, and why are we so sure that harm would ensue? While theologians cited religious traditions and biblical proscriptions, lawyers cited reproductive rights and said it would be very hard to argue that it was illegal to clone oneself. In the meantime, some ethicists said they'd heard from in vitro fertilization clinics, which—operating already outside the usual rules that bind scientists, and looking for paying customers—were extremely interested in investigating cloning.

The diversity of opinions extended even to interpretations of identical passages from the Bible. One priest and Catholic theologian argued from Genesis that cloning would be against God's will. An orthodox rabbi and theologian argued from the same passage that cloning should not be proscribed.

The priest, Albert Moraczewski, of the National Conference of Catholic Bishops, was invited to explain the Catholic point of view by a presidential commission that was asked to make recommendations on whether cloning should be permitted. He began by saying that the cloning of humans would be an affront to human dignity. Then he spoke of the familiar story of Adam and Eve, told in the Book of Genesis, in which God gave humans dominion "over the creatures that swim in the sea, that fly in the air, or that walk the earth." And he spoke of God's order. "The Lord God gave man this order: 'You are free to eat from any of the trees of the garden except the tree of knowledge of good and bad.' "

Moraczewski explained that according to the Catholic interpretation, "Adam and Eve were given freedom in the garden but with one limitation, which if transgressed would lead to death. Accordingly, human beings have been granted intelligence and free will so that human beings can search for, and recognize, the truth and freely pursue the good."

Cloning, he said, would exceed "the limits of the delegated dominion given to the human race. There is no evidence that humans were given the power to alter their nature or the manner in which they come into existence."

He added that couples who clone a child would be dehumanizing the act of procreating and treating their child as an object, attempting to "design and control the very identity of the child."

Moraczewski concluded by quoting John Paul II: "The biological nature of every person is untouchable."

The next day, Moshe Tendler, an Orthodox Jewish rabbi, spoke to the commission. He, too, started with Genesis, and with the same quotation. But his interpretation of it, from the Jewish tradition, was very different.

"This knowledge of good and evil has always confused theologians and certainly the layman," Tendler said. "If Adam and Eve did not know of good and evil, how could they have sinned? They knew good and evil. The tree of good and evil is the tree that allows you to think that you can reevaluate, you can set another yardstick for what is good and what is evil."

The Jewish tradition says that humans are obliged to help master our world, according to Tendler, as long as they do not transgress into areas where they would attempt to contravene God. It would not be in character with the Jewish tradition to have a technology that could have outcomes that are good—like preserving the family line of a Holocaust survivor who had no other living relatives—and decide, ahead of time, not to use it for fear of its evil consequences. "We are bound by good and evil as given to us by divine imperative. And we knew pretty well in most areas what is good and what is evil until cloning came along and now we are not so sure what is good and what is evil.

"So, cloning, it is not intrinsically good or evil," Tendler said.

The question, really, is whether particular applications of cloning might be a transgression by humans into the domain of God.

"I will give you a simile or metaphor of a guest invited to your house," Tendler said. "You ask them to be comfortable, help themselves, there is cake in the cake box and fruits in the refrigerator, and coffee in the coffeemaker." When you wake up, he continued, you're pleased to see that your guest did as you suggested. "But if he should move your sofa to the other side of the wall because he thought that that is where it really belongs, you will not invite him again."

God, Tendler added, says, "Make yourselves comfortable in my world, but you are guests in my house, do not act as if you own the place. Don't you rearrange my furniture."

He spoke also of a metaphor from the Talmud. "The question was posed, 'Is there not a time when you say to the bee, neither your honey nor your sting'?" And so, he asked, are we really prepared to ban cloning, to give up the honey, because we are so afraid of the sting?

On the other hand, some wonder whether we might not want to squash the bee. Nancy Duff, a theologian at the Princeton Theological Seminary, argued from Protestant tradition that, at the very least, all thoughts of human cloning should be put on hold. "Many people wonder if this is a miracle for which we can thank God, or an ominous new way to play God ourselves," she said. "At the very least, it represents the ongoing tension between faith and science."

But there is also a secular point of view, one that asks how persuasive, after all, are the hypothetical harms of cloning, and whether they are great enough to override the right that people have to reproductive freedom. John Robertson, a law professor at the University of Texas in Austin, who specializes in ethics and reproductive law, said he is unconvinced by those who argue that cloning is somehow too unnatural, too repugnant, too contrary to the laws of God, to proceed with. "In assessing harm, deviation

from traditional methods of reproduction, including genetic selection of offspring characteristics, is not in itself a compelling reason
for restriction when tangible harm to others is not present." He
argued that cloning is not significantly different from other methods our society now accepts as ethical, and which are now being
actively studied in research laboratories throughout the world. He
referred to methods for adding genes or correcting faulty ones, in
an attempt to cure diseases like muscular dystrophy or cystic
fibrosis, which, although not yet possible, is expected to occur
before too long.

"Cloning enables a child with the genome of another embryo
or person to be born," Robertson said. "The genome is taken as
it is. Genetic alteration, on the other hand, will change the genome of a person who could have been born with their genome
intact." So what is the greater intervention? Given a choice of a
child who is a clone or no child at all—a choice that could befall
infertile couples—how bad is it to allow them to have a clone?
Robertson asked. "If a loving family will rear the child, it is
difficult to see why cloning for genetic selection is per se
unacceptable."

A compelling argument, said Daniel Brock, a philosopher and
ethicist at Brown University. Is the right to clone part of our
right to reproductive freedom? he asked. He said that although
he is not certain that cloning could be protected in this way
because it is not, strictly speaking, reproduction, it might nonetheless fall into that broad category. And, he added, if the right
to have yourself cloned is treated as a reproductive right, "that
creates the presumption that it should be available to people who
want to use it without government control."

Brock, for one, thinks that the public reaction to cloning is
overblown. "The various harms are usually speculative," he said.
"It is difficult to make the claim that these harms are serious
enough and well-enough established to justify overriding the claim
that cloning should be available." The public, he said, "has a

tendency to want to leap ahead to possibilities that we're not even sure are possible."

Ruth Macklin, an ethicist at Albert Einstein College of Medicine, raised similar questions about whether fears of cloning are reasonable. "One incontestable ethical requirement is that no adult person should be cloned without his or her consent," Macklin said. "But if adult persons sought to have themselves cloned, would the resulting individual be harmed by being brought into existence in this way? One harm that some envisage is psychological or emotional distress to a person who is an exact replica of another. Some commentators have elevated this imagined harm to the level of a right: the right to control our own individual genetic identity. But it is not at all clear why the deliberate creation of an individual who is genetically identical to another living being (but separated in time) would violate anyone's rights."

After all, Macklin said, if the cloned person was not created from the cell of another, he or she would not have been born. Is it really better never to have existed than to exist as a clone? "Evidence, not mere surmise, is required to conclude that the psychological burdens of knowing that one was cloned would be of such magnitude that they would outweigh the benefits of life itself."

Macklin even took on those who argued that cloning violates human dignity. Those who hold that view, she said, "owe us a more precise account of just what constitutes a violation of human dignity if no individuals are harmed and no one's rights are violated. Dignity is a fuzzy concept and appeals to dignity are often used to substitute for empirical evidence that is lacking or sound arguments that cannot be mustered."

Kass argued, however, that such utterly pragmatic language obscures the moral significance of what is being contemplated. He quoted Bertrand Russell: "Pragmatism is like that warm bath that heats up so imperceptibly that you don't know when to scream."

The clashing viewpoints, said Ezekiel J. Emanuel, a doctor and ethicist at the Dana-Farber Cancer Institute in Boston, who was a member of the president's commission that was studying cloning, seem to indicate "a moral values gap." And so, he added, how people react to cloning "depends a lot on one's world outlook, as it were. How much you might weigh these other values depends a lot on how you understand yourself and your place in the world."

And that, in the end, is what cloning brings to the fore. Cloning is a metaphor and a mirror. It allows us to look at ourselves and our values and to decide what is important to us, and why.

It also reflects the place of science in our world. Do we see science as a threat or a promise? Are scientists sages or villains? Have scientists changed over the years from natural philosophers to technologists focused on the next trick that can be played on nature?

Freud once said that, sometimes, a cigar is just a cigar. But so far, we have not reached a point where a clone is just a clone. As the social and cultural history of cloning continues, the questions and the insights into who we are, who we are becoming, and who we want to be grow ever deeper. Dolly, it now seems, is more a beginning than an end.

2

BREAKING THE NEWS

You mean, given that it's the most important story of the
last two or three decades?
                    —JAMES FALLOWS
              Editor, *U.S. News & World Report*

Perhaps the most surprising parts of the cloning story were its
origins and its guiding forces. It was the story of our age: The
practical scientist finds a sponsor and does the unthinkable. A
scientific journal tries its best to keep the story under wraps to
ensure itself the maximum publicity. The media go into a frenzy,
combing the country for any expert, anyone who has had fifteen
minutes of fame on the cloning question, searching for new angles
on the event, almost oblivious to the larger picture and the historical context.

In many ways, it is a story that could not have happened at
any time but now.

Even though scientists were shocked by Wilmut's feat, even
though Dolly's birth seemed to come out of nowhere, Dolly was
long in the making, the product of a decade-long research project
quietly directed by the eminently practical Ian Wilmut. Wilmut

was fifty-two years old when Dolly was born, an embryologist with impeccable credentials. With his neatly cropped auburn beard and gentle face, he was the antithesis of the mad scientist. By the time Dolly was born, he had worked at the Roslin Institute for twenty-three years, laboring for nine hours a day, leaving the lab at six each night and, more often than not, bringing work home. The cloning work was long and tedious. It required infinite patience and an ability to work long hours hunched over a microscope in a tiny room heated to the internal temperature of a sheep. It was work that few cared about and that almost no one wanted to fund. And so, perhaps, it was even more remarkable that Wilmut and a handful of his colleagues persisted.

Wilmut is an Englishman, born in Hampton Lucey and educated at the University of Nottingham. There he discovered embryology, studying with G. Eric Lamming, a world-renowned expert. He entered Darwin College in Cambridge in 1971 and received a Ph.D. two years later—remarkably fast compared to the four to eight years many scientists take to pass their graduate-level courses, take written and oral exams in their subjects, and do the original research that is necessary.

He married and had three children: Helen, who was twenty-eight when Dolly was announced; Naomi, who was twenty-six; and Dean, who was twenty-four. And he settled in for the long haul at the Animal Breeding Research Station in Scotland, which became the Roslin Institute.

It was a bucolic life in a tiny town just south of Edinburgh that drew few visitors. There is a scattering of castles nearby. The Elizabethan poet William Drummond lived in one, Hawthornden, which is now a writers' retreat. And there is the Roslin Chapel, known for its architecture. But most people had no reason to know about, or visit, the town.

Wilmut's hobby is walking in the mountains near his home and his favorite drink is single-malt Scotch. Although his wife, Vivian, is an elder in the Church of Scotland, Wilmut is not religious. "I don't have a religious belief," he said. "I consider myself an agnostic." He is quiet, modest, and as the world reacted to the birth of Dolly, a bit startled to have been thrust onto center stage. He knew, of course, that what he had done was important. "I am not a fool," he said. "I know what is bothering people about this. I understand why the world is suddenly at my door. But this is my work. It has always been my work, and it doesn't have anything to do with creating copies of human beings. I am not haunted by what I do, if that is what you want to know. I sleep very well at night." For Wilmut, the goal was to develop animals that could produce drugs for human use.

Scientists have long known what was involved in trying to clone, and many had convinced themselves that it was biologically impossible. The problem begins with the mysteries of embryo development. Every cell in the body arises from the same fertilized egg and so every cell in the body has exactly the same genes. But animal—and human—cells are specialized, differentiated, so that a heart cell behaves like a heart cell and a liver cell like a liver cell. This process of differentiation begins almost as soon as a fetus forms, and once a cell has reached its final state, it never alters. A brain cell remains a brain cell for as long as a person is alive; it never turns into a liver cell, even though its genes are the same.

One key to differentiation lies in the proteins that coat the DNA in the nucleus of a cell. These proteins mask as many as 90 percent of a cell's genes, leaving open only those that the cell needs to survive and to perform its specialized functions. To clone, investigators would have to entice the DNA of a specialized cell to lose the proteins that bind to its DNA and take up those that bind to the DNA of a newly fertilized egg. It is impossible for biologists to simply strip the proteins off a cell's chromosomes in order to reprogram a cell. Chemicals that might tear the proteins off are too harsh—they would shatter the DNA, destroying it.

So the question for erstwhile cloners was, Is it possible to erase the ravages of time on a cell's genes—to return a cell to its original undifferentiated state—and allow the cell to direct the development of an entire new organism?

Wilmut's feat, the cloning of a lamb from an adult udder cell, was, therefore, no ordinary one. To biologists, it was like the breaking of the sound barrier or, perhaps more appropriately, the splitting of the atom.

Yet Wilmut was remarkably taciturn about the event. Even the process by which he selected the type of cell to clone seemed eminently pragmatic. Wilmut cloned Dolly from an udder cell of a six-year-old lamb, using cells that were simply convenient— they happened to be stored in a laboratory freezer and of interest to Wilmut's sponsor, PPL Therapeutics, Ltd., a tiny biotechnology company with headquarters in Edinburgh.

PPL wanted to use sheep to make drugs to treat human diseases like hemophilia. The company, whose name originally stood for Pharmaceutical Proteins, was formed to commercialize research under way at the Roslin Institute. Scientists there had found that they could genetically engineer sheep so that the animals made a drug, alpha-1 antitrypsin, in their milk. It was not a very efficient process, but it could be economically feasible since the drug, which can be used to treat cystic fibrosis, was so valuable.

But the real payoff, the company realized, might be from cloned animals. Cloned sheep could become living drug factories and might produce valuable drugs much more cheaply than did the methods used by drug companies.

The idea behind cloning a sheep so it made drugs was fairly simple. Scientists would take sheep cells and grow them in the laboratory. There, they would add genes to the cells. Genes direct cells to make specific proteins that in this case would also be drugs. Scientists might, for example, give the cells genes that directed them to make fibrinogen, a protein that is also a drug that can help in wound healing.

So far, no one has found a good way to coax cells to take up added genes and start using them to make proteins. Typically, about one cell in a million does what the scientists want. But that is of little concern to cloners because when the scientists are growing millions of cells in a laboratory, all they have to do is to fish out the few cells that have accepted their gene and are making the drug and use them in the next steps of their experiment.

With the drug-producing cells in hand, the scientists would then be ready to clone. They would clone a lamb whose udder cells made the drug whenever they made milk—all they'd have to do is hook the drug-producing gene to the gene that is turned on when milk is produced and make clones from those genetically altered cells. Then the company could simply milk the sheep, extract the drug from the milk, and sell it. If the scientists made both male and female sheep that carried the added gene, they could breed these sheep and have a self-perpetuating flock of living drug factories.

"What this mostly will be used for is to create more health-care products," Wilmut told me soon after the cloning of Dolly was announced. He said he started with sheep because "they are cheaper than cows," which produce far more milk.

Wilmut also envisioned creating cloned animals that would serve as models for human diseases. For example, he could give sheep a gene for cystic fibrosis, using the same genetic engineering methods he hopes to use to clone sheep that make drugs. Sheep with the cystic fibrosis gene could be used to test new treatments, even gene therapy, that scientists are devising for humans. And they could be used in basic research studies of how the genetic defect causes the deadly disease.

In addition, Wilmut said, he'd like to use cloning to study scrapie, a degenerative neurological disease of sheep that resembles mad cow disease. British cattle are thought to have gotten mad cow disease from eating feed made from the remnants of sheep. And many fear that people have gotten the disease from eating

British beef. But scrapie is a poorly understood disease. One idea is to clone sheep with a genetic defect that is thought to predispose animals to scrapie in order to study how and why the disease develops.

To clone, Wilmut used methods his research group and others had been developing for more than a decade. His colleague Keith Campbell sucked the nucleus out of an egg that had been removed from a ewe, creating an egg that had no genes at all, an egg that would soon die if it did not get a new nucleus. Then he began the process of adding the nucleus of an udder cell to the bereft egg.

Campbell slipped an udder cell under the outer membrane of the egg. Next, he jolted the egg for a few microseconds with a burst of electricity. This opened the pores of the egg and the udder cell so that the contents of the udder cell, including its chromosomes, oozed into the egg and took up residence there. Now the egg had a nucleus—the nucleus of the udder cell. In addition, the electric current tricked the egg into behaving as if it were newly fertilized, jump-starting it into action. After 277 attempts to clone an udder cell, Wilmut's group succeeded and Dolly was created.

The sheep that provided the udder cell, ironically, is lost to history. "She was put down," Wilmut explained to me about two weeks after the Dolly story broke, over lunch in a Baltimore hotel, where he was speaking to rapt animal scientists. The sheep was actually living on another farm, some distance from Wilmut's lab, and, at age six, it was time for slaughter. Some unknowing persons butchered her and sold her. The sheep was eaten. Her udder cells were merely cells of convenience, Wilmut explained. PPL Therapeutics happened to have the cells frozen in its laboratory, as part of its research projects, so Wilmut used them.

Wilmut, however, had evidence that Dolly was indeed a clone. He had frozen some of the original udder cells so that when Dolly was born he could show, with DNA fingerprinting, that Dolly's genes were the same as those of the udder cells and that they bore no resemblance to those of the white sheep that provided the egg or the Blackface sheep that served as Dolly's surrogate mother.

After Dolly was born and appeared healthy, and before she could be introduced to an incredulous world, there were practical matters to be disposed of. PPL Therapeutics wanted a patent on the use of the cloning technology to produce animals that made drugs in their milk. So Wilmut and his ten collaborators in the cloning research kept mum. Months later, they wrote a four-page scientific paper and submitted it to *Nature,* a British science journal. It was accepted on January 10 and scheduled to be published on Thursday, February 27, 1997. And until it was published, the plan was for no one to breathe a word.

Here again, the tyranny of the mundane came into play. A lamb had been cloned, and it was kept secret because of a patent application. And when the announcement was nigh, the quotidian jostlings of journals and journalists erupted. There was a sort of game under way between the science journal, intent on getting maximum publicity for its paper, and journalists, intent on getting maximum publicity for themselves and their newspapers or television shows, which made the feat into something of a comic opera.

Nature is one of a handful of science journals that are published weekly. It is one of the more prestigious journals, rejecting many more papers than it can publish. And it has a long history of publishing major papers. So most scientists avidly scan the journal each week, as do the world's journalists. A paper in *Nature* comes with an imprimatur of approval and a promise of importance.

Like its competitors, *Nature* attempts to get the maximum attention for its articles. The goal of these journals is to have virtually every newspaper and television show in the country reporting on one or more of its articles on the very day the journal is published. Like other journals, *Nature* gives reporters a head start on its articles. It sends out a tip sheet to journalists containing short descriptions of its articles for the forthcoming issue and will provide faxes of entire articles upon request. *Nature* tip sheets arrive a week before the journal is published, in reporters' E-mail boxes. The next day, reporters can get faxed papers. In return, the journalists promise not to publish or broadcast stories about *Nature* papers until the day *Nature* is published.

On Thursday afternoon, February 20, the *Nature* E-mail arrived in my mailbox with a brief statement: "The lamb on this week's cover was raised from a single oocyte (egg cell), whose nucleus had been replaced with that from an adult sheep mammary gland cell. It may be the first mammal to have been raised from a cell derived from adult tissue." It concluded, with historic understatement, "The implications of this work are far-reaching."

The paper itself had a title that said nothing about clones. It was called "Viable Offspring Derived from Fetal and Adult Mammalian Cells." And it conformed to the stylized form of scientific writing that is as rigid as a haiku. It began, as such papers always begin, by nodding to predecessors. It concluded, as such papers always do, by noting what the results might mean. But never did Wilmut spell out, in full detail, what his result implied to the world.

The paper began by saying that investigators have long been able to start with frog skin cells and clone a frog embryo that grows to the tadpole stage. But, invariably, these cloned frog embryos died when they got to the tadpole stage. And so, Wilmut wrote, it has remained uncertain whether it is possible to clone an adult cell so that the resulting embryo grows normally and does not die at an early stage of development. Scientists had questioned whether it was possible to reprogram the genes of an adult cell, fearing that they were too specialized to revert to the state when a sperm first fertilizes an egg. All this was said, however, in two terse sentences: "It has long been known that in amphibians, nuclei transferred from adult keratinocytes established in culture support development to the juvenile, tadpole stage. Although this involves differentiation into complex tissues and organs, no development to the adult stage was reported, leaving open the question of whether a differentiated adult nucleus can be fully reprogrammed."

Wilmut's paper continued, stating the goal of the experiment. He proposed that he might be able to clone an adult cell if he first put the cell into a resting phase so that it was not actively getting ready to copy its DNA and divide. He wrote: "Here we investigate whether normal development is possible when donor cells derived from fetal or adult tissue are induced to exit the growth cycle and enter the G0 phase of the cell cycle before gene transfer."

After describing his method and his result, the birth of eight live lambs, one of which came from a cloned adult cell and the rest from cloned specialized fetal cells or unspecialized sheep embryo cells, Wilmut came to his conclusion: that cloning should be useful in the livestock industry by allowing breeders to clone herds with the characteristics of prize adults, like cows that are record-breaking milk producers. It should be useful in biotechnology because it will allow scientists to grow cells in a laboratory, add genes that interest them, like genes that cause the cells to produce drugs, and then use those cells to create cloned animals. And the technique should be useful to researchers who want to understand changes that occur in DNA when eggs are first fertilized and changes that occur during development and aging. Nowhere did Wilmut mention ethical quandaries.

He wrote: "Together, these results indicate that nuclei from a wide range of cell types should prove to be totipotent after enhancing opportunities for reprogramming by using appropriate combinations of cell-cycle stages." And, he added in the same abstruse language, these processes should make it possible to clone animals with desired genetic traits "by nuclear transfer from modified cell populations and will offer new opportunities in biotechnology." Finally, Wilmut wrote, cloning could allow scientists to "study the possible persistence and impact of epigenetic changes, such as imprinting and telomere shortening, which is known to occur in somatic cells during development and senescence, respectively."

Nature had also commissioned an editorial by Colin Stewart, an embryologist at the National Cancer Institute in Frederick, Maryland, to accompany Wilmut's paper. Continuing in the careful tone of the tip sheet and of Wilmut's paper, it focused only on the cloning of farm animals, and even had the jokey title, "An Udder Way of Making Lambs." In jargon-filled technical writing, Stewart essentially avoided the earthshaking significance of the work. After discussing how the cloning was done and what science it was built upon, Stewart came to conclusions similar to Wilmut's, expressed in the most impenetrable of scientific jargon, concluding on a note of whimsy, "Maybe, in the future, the collective noun for sheep will no longer be a flock—but a clone."

The cover of the issue of *Nature* was sky blue, with a photo of an aquamarine petri dish in its center. Superimposed on the petri dish was a lamb. In white letters was written underneath, "A flock of clones."

But scientists were not fooled by the dense language and narrow focus, and neither were some journalists. I saw the E-mailed tip sheet almost as soon as it was sent, on Thursday, February 20. I was working at home, as I often do, so I called the deputy science editor at *The New York Times,* Jim Gorman, and told him that if the tip sheet was not exaggerating, *Nature* was going to publish an astounding paper on the first cloning of an animal. I said that I'd get the *Nature* paper on Friday, as soon as it was available, and tell him what it said.

On Friday, I got the paper and saw that Wilmut really did claim to have cloned a lamb from an udder cell of an adult sheep. Jim Gorman and I decided that news so important was unlikely to wait for the usual *Nature* embargo to end. In journalism, the rule is that once a newspaper, television show, or radio show reports on an embargo story, it is fair game for everyone to break the embargo. We decided that I would get a major story ready to go and that the *Times*'s editors would continually check the wire—the stories that come into our computers by special arrangements with news services like the Associated Press and Reuters—and that should alert us immediately if another news organization had reported *Nature*'s cloning story. If and when that happened, the *Times* would rush my story into print.

I wrote the story and also, at the suggestion of Cory Dean, the science editor, I included a box answering all the questions we could think of on cloning: Could you clone the dead? Who are the parents of a clone? Is it legal to clone people? (The answers are, No you can't clone the dead because cloning, as Wilmut does it, involves the fusion of a living cell with an egg. The genetic parents of a clone are the man and woman whose sperm and egg formed the person who, in turn, was cloned. The legal parent is the woman who bears the cloned baby. As of this writing, it is illegal to clone people in Britain, Spain, Denmark, Germany, and Australia but not in the United States.)

As Saturday, February 22, wore on, and the cloning story had not broken, the editors of *The New York Times* and I thought that the embargo probably would last through the weekend. I went out on Saturday night. When I returned, around midnight, there were five messages on my answering machine, all from editors at the *Times,* all asking me to call immediately. It turned out that Robin McKie, the science editor of *The Observer,* a British paper, had published a story about Wilmut's feat. McKie later revealed that he relied on his own sources for the story so he did not actually break *Nature*'s embargo. But the effect was the same.

The New York Times immediately went into action, publishing its story in time for the second edition of the paper, which meant that only those readers who lived closest to New York City and got the edition of the Sunday paper that was printed late Saturday afternoon could have been unaware of the cloning of Dolly. It was printed on the left-hand side of page 1 in the second edition, making it the

second most important story in the paper, after a story titled DEMO-
CRATS SHOW SOME FLEXIBILITY ON CAPITAL GAINS. By the next edition,
the editors had made cloning the lead story, placing it on the right-
hand side of the page, signaling that they thought this was a major
event. The rest of the press followed suit, scrambling to get out their
own page 1 stories on Monday. Meanwhile, I wrote a second page
1 story that was published in the *Times* on Monday, spelling out the
theological and ethical implications of cloning.

Time and *Newsweek* put cloning on their covers the week after
it occurred. *U.S. News & World Report,* their competitor, did not,
and the magazine's failure to do so was explored in an article in
The New Yorker that was a profile of the magazine's editor, James
Fallows. The article said that the only questions for *Time* and
Newsweek were "whether to use images of sheep (*Time*) or people
(Wilmut) on the cover and how large a proportion of each maga-
zine the cloning story would consume."

But Fallows was out of step. His boss, Mort Zuckerman, wanted
cloning on the cover. Fallows insisted on his previous choice, a
story on "America's Best Graduate Schools." He said he felt his
hands were tied because the story on graduate schools was timed
to the publication of a paperback book, which had already been
shipped to stores. "It would look odd to have the book go on
sale ten days before the magazine," Fallows said.

Zuckerman tried to compromise, suggesting, for example, a
cloning cover for subscribers and a graduate-school cover for the
newsstands. But in the end, the magazine stayed with its graduate-
school cover for all its editions. Yet, ironically, the graduate school
book was recalled before it could be sold because *U.S. News* had
miscalculated its rankings of law schools.

The *New Yorker* writer asked Fallows "if he was upset that
logistics prevented him from putting cloning on the cover. He
smiled and said, 'You mean, given that it's the most important
story of the last two or three decades? Yes.' "

*   *   *

Soon Wilmut and the tiny village of Roslin, where there are more sheep than people, were caught up in a media extravaganza. The press descended upon the town, hounding the scientists, camping out, insisting on seeing Dolly, on interviewing Wilmut. One Greek television reporter demanded to be allowed to leap into the pen with Dolly to show her audience that the lamb really existed. She did not get permission.

Wilmut told me he immediately got more than four hundred E-mail messages from around the world, many from school-children. He divided them into five groups, devised a stock response to the messages of each group, and answered them all.

A week after the media swarm overtook the Roslin Institute, desperate scientists declared a "media-free day." Researchers locked themselves in their labs. Dolly was hidden. But, of course, by then it was too late. Cloning fever had hit the world and no expert, no source of information, went untapped. At the University of Pennsylvania, Arthur Caplan, the director of the university's Center for Bioethics, watched aghast as his center's Internet site became jammed. Before Dolly, the site was getting about five hundred hits a month. After Dolly, he told me, it began getting seventeen thousand hits a day.

Cloning jokes began making the rounds, with favorites playing off William Blake's famous line, "Little lamb, who made thee?" One reply: a significant udder.

Canon, the company that makes photocopiers, came out with an advertisement featuring two identical sheep. "Big deal. We've been making perfect copies for years," its ad said.

Religious groups and the Vatican weighed in, saying that cloning of humans would be abhorrent, the ultimate in hubris, an attempt by humans to be their own creators. The World Health Organization said it opposed the cloning of humans. The Humane Society opposed the cloning of animals. The American Society of Reproductive Medicine issued a press release saying it "finds the practice of human cloning through nuclear transplantation unacceptable."

But not every organization was opposed to the cloning of humans. In New York, a pro-cloning group sprang up, the Clone Rights United Front, whose members included gay men and lesbians who wanted to clone themselves. The lesbians envisioned taking a cell from one woman and implanting it in an egg from another, thus creating a baby without the presence of a man.

Scientists who had been known only to other specialists were suddenly in demand as reporters discovered they could provide pithy quotes. Lee Silver, for example, a molecular biologist at Princeton University, who had been all but unknown to the media until he was quoted in a newspaper story, turned up on twenty-three talk shows in the first two weeks after Wilmut's feat was reported, and he testified before the New York State Senate.

President Bill Clinton wasted no time in acting. By sheer good luck, he had appointed a national bioethics commission several months before to advise him on the ethical problems that may arise in medical research that uses human subjects and to examine the issues of genetic information and how it is handled by medical institutions. Now he already had in place just what he needed to evaluate cloning: an august group of experts chaired by Harold Shapiro, the president of Princeton University.

On Monday, February 24, Clinton wrote to Shapiro, asking that the commission swing into action. The cloning of Dolly, he wrote, "raises serious ethical questions, particularly with respect to the possible use of this technology to clone human embryos." And so, he continued, "I request that the National Bioethics Advisory Commission undertake a thorough review of the legal and ethical issues associated with the use of this technology, and report back to me within ninety days with recommendations on possible federal actions to prevent its abuse."

In Europe, Jacques Santer, head of the European Commission,

requested an opinion from a group of nine experts on science, the law, philosophy, and theology, on the ethical consequences of cloning. The timetable was the same as for the American commission—an opinion was expected in May of 1997.

The next week, the House held hearings on cloning. The following week, the Senate followed suit. Harold Varmus, a Nobel laureate who directs the National Institutes of Health, testified at both hearings, telling the senators and representatives that he opposed laws to ban cloning because it would be impossible to craft a law that might not inadvertently cut off valuable research that could save lives. The Senate hearing, on March 12, featured Wilmut himself, who had come to the United States to speak to an animal biotechnology group in Baltimore on March 10 and who was snared by the Senate while he was in the country.

The senators seemed awestruck. Christopher Dodd of Connecticut groped for an appropriate analogy. "I'm trying to think of events in my own lifetime" that would be comparable to the cloning of Dolly, he said. "I suppose the dawn of the nuclear age. Here we seem to be entering the genetic age."

"I think you are one of the true trailblazers in human history," Senator Tom Harkin of Iowa told Wilmut.

The cloning, said Senator Bill Frist of Tennessee, a heart surgeon and the committee chairman, was "a real breakthrough, not just an evolutionary development in science." He added, "It challenges our imagination and beliefs about the nature of knowledge."

Wilmut seemed a bit dazed by the attention. "I have to admit that the sheer scale of it has taken me by surprise," he said.

Even many scientists, who often are jaded or professionally cautious, were taken aback. Marie Antoinette Di Berardino, a serious woman who was a cloning pioneer, and who had been studying cloning in frogs since the 1950s, said she could imagine nothing that would justify cloning a human being. "I'm coming from a

Roman Catholic background, and I'm telling you, it's killing me,"
she said. "We have lost so much respect for human life in so
many ways. I just have a fear that this may be taking a step
too far."

Lee Silver was stunned when he learned of the cloning. "It's
unbelievable," he said. "It basically means that there are no limits.
It means that all of science fiction is true. They said it could
never be done and now here it is, done before the year 2000."
Silver was unable to sleep the night he first learned about the
cloning of Dolly. It was a historic moment, he realized, and time
for him was now divided into pre-Dolly and post-Dolly.

R. Alta Charo, a lawyer and ethicist at the University of Wisconsin, said that cloning came at a time when scientists and
laypeople were increasingly convinced that genes are destiny, and
so the idea of cloning genetically identical individuals seemed
terrifying.

"I think that we go through historic waves of emphasis on
nature versus nurture in our gut instinct over what will largely
determine a person's place in the world, as well as their personality, character, instincts, and hobbies," she said. And, she added,
"We are very much in a biological deterministic wave right now.
Every week, we see reports of another gene that has been mapped
and associated with behavior or disease. It feeds our notion that
we are reacting to programmed instructions from our genes and
that leaves little room for free will."

Scientists, of course, are not naive, Charo said. They understand
that individuals are formed by still poorly understood and unpredictable interactions of their genes with their environment. Yet
at the same time, scientists are giving laypeople a very mixed
message. At the very moment that they are telling people not to
overestimate the importance of genes, they are telling them how
important genes are. "It's a very complicated set of messages to
hold in one's mind simultaneously, and most people are too busy

with their regular workaday lives to spend time to carefully evaluate what they read and to carefully put it in context," Charo said. "That's just not realistic."

At the same time that we believe in our hearts that genes are destiny, Charo noted, "we celebrate our individuality—we celebrate it to the nth degree." And so, "in an era where it is harder and harder to believe that physical uniqueness will exist if our genetic substrate is identical, the idea of having people who are not socially unique is associated with lots of our nightmares. The most frightening footages from World War II are not the ones that show the skeletal bodies of the survivors and not even the mushroom clouds over Hiroshima and Nagasaki. The most frightening are the crowds of Germans, with their fists raised, shouting Sieg Heil. The most frightening is the mob psychology that reduces individuals to nothing but clones of each other, with an absence of questioning authority or forming independent opinions. It is associated with our fear of people who can be easily manipulated, who can become an unthinking mass that can be a force of oppression." And so, Charo said, "as long as we are in a biologically essentialist era, that fear is going to be hard to fight." That fear is at the heart of our fear of cloning.

Some humanists, however, seemed almost blithely indifferent, shrugging off the cloning as simply too distant to contemplate, and remarking that cloning, even the cloning of humans, would not be the end of humanity as we know it.

Joyce Carol Oates, the novelist, short-story writer, and professor at Princeton University, said she was not so concerned by cloning. At a Princeton wedding a few weeks after Wilmut's cloning feat had been announced, standing on the fringes of a noisy crowd, Oates asked me what it was people were so afraid of. The possibility that the cloning of humans might be a threat to our individuality? "Most people don't have any individuality anyway," she said to me.

"There is a brilliant analogy by one of the great Talmudists,"

Moshe Tendler said. The scholar tells us that "the sword and the book came down from heaven intertwined." God said, "Choose the book and live or the sword and die." But, Tendler said, the Talmudic interpretation is that God did not say the sword *or* the book. God gave us the sword and the book together. So, Tendler concluded, the question is "Are you prepared to sheathe the sword with the book or are you willing to give up the book and have a naked sword?"

"All human activity has a potential for good and bad," Tendler said. "Only because of our ethical and moral sense can we direct our activities so that the bad is hidden by the good. The job of man is to make sure the sword is always sheathed by the message of the book."

Others feared for the future of humankind. John Paris, a small, white-haired Jesuit priest from Boston, told me he is certain now that humans will be cloned. "I cannot imagine a world in which someone won't try it," he said. "There are two things that drive us—power and money. And fame leads to fortune. Somebody will try it."

Those who want to clone, said Stanley Hauerwas, a divinity professor at Duke University, "are going to sell it with wonderful benefits" for medicine and animal husbandry. "They will say, 'We'll start with animals and then we can do all kinds of things with wonderful therapeutic possibilities.' " But his concern is with "the profounder issue," one involving the hubris of humans.

"I do think there is a kind of drive behind this for us to be our own creators," Dr. Hauerwas said. Why have God, he asked, when we can create ourselves, when we can "get out of this life alive" by creating a clone of ourselves to live on after us?

George Annas, a rotund law professor at Boston University, with a short bushy beard, begged for a law to prohibit the cloning of a human being. "It's not just another infertility treatment. It should be viewed with horror," he told the senators at Bill Frist's hearing. And so he pleaded for legislation. "We know where we're

going and now we can ask—for one of the few times in history—
do we want to go there?"

But others found it hard to imagine enforcing a ban on cloning
people when cloning gets more efficient. "I could see it going on
surreptitiously," said Lori Andrews, a professor at Chicago-Kent
College of Law. For example, in the early days of in vitro fertilization, Australia banned the practice. "So scientists moved to Singa-
pore" and offered the procedure.

"I can imagine new crimes," Andrews said. People might be
cloned without their knowledge or consent. After all, all that
would be needed is some cells. If there is a market for a sperm
bank selling semen from Nobel laureates, how much better to
bear a child that is actually a clone of the great thinker or perhaps
a great beauty or great athlete that you admire.

"The genie is out of the bottle," said Dr. Ronald Munson, an
ethicist at the University of Missouri. "This technology is not, in
principle, policeable."

And, he added, the future possibilities are incredible. For example, could researchers devise ways to add just the DNA of an
adult cell to an egg whose nucleus was removed, without fusing
two living cells? If so, might it be possible to clone the dead?

"I had an idea for a story once," Munson said, in which a
scientist obtains a spot of blood from the True Cross, where Jesus
was crucified. He then uses it to clone a man who is Jesus Christ—
or perhaps not, perhaps he couldn't be.

Inevitably, perhaps, some scientists began to downplay Wilmut's feat. It is only sheep, they said, and the cloning process is
difficult and unreliable. After all, it took 277 tries to get Dolly.
Would it be economically feasible even to clone farm animals,
like prize milk cows, with such a system?

But others said that these scientists were being perhaps purposely obtuse. "It is so typical for scientists to say they are not
thinking about implications, to say, 'We're just doing our own
little thing,' " Silver told me. "This is what scientists have been

doing forever. The scientists who do the research never think about the implications. On the surface, the only way they can validate what they are doing is to say they are only doing it in sheep. If something can be extrapolated to humans, other people can do it, not us. We don't want to be involved. It might affect our ability to do research." It can be impossible to do research if you think about where it might lead. Scientists can become paralyzed. So, Silver said, some have decided simply not to think about the larger issues.

Alan Weisbard, the ethicist at the University of Wisconsin, said that there also are "institutional pressures" on scientists and policy makers to "calm down public speculation and play down certain speculative possibilities." Weisbard disagreed with that approach. It risks losing "opportunities for public education and public thought," he explained. Yes, cloning is frightening, and yes, it is provocative. But, he said, "it is time for us as a society to understand that the answers do not come so quickly."

For Wilmut, and for some other scientists, the cloning of humans may be provocative, but it is not what is foremost in their minds. Their eyes are fixed on the practical, and on the good that cloning can do.

So the magnificent and chilling experiment, the creation of the sheep that changed the world, was perhaps science as only dedicated scientists could conceive it. Ronald Munson captured the irony.

"Here we have this incredible technical accomplishment, and what motivated it? The desire for more sheep milk of a certain type." It is, he said, "the theater of the absurd acted out by scientists."

3

NATURAL PHILOSOPHIES

Decisive information about this question may perhaps be afforded by an experiment which appears, at first sight, to be somewhat fantastical.

—Hans Spemann,
1938

If the actual cloning of a lamb was an act of utter pragmatism, the work leading up to it came from a tradition of utter romanticism. Long before anyone knew the nature of genes, long before there was such thing as molecular biology, biologists had thought of cloning. The idea came not because mad scientists wanted to create hundreds of identical twins of animals or people, but because cloning turned out to be deeply tied to their efforts to understand the abiding biological mysteries of development and the psychological mysteries of identity.

Ever since the time of the ancient Greeks, philosophers and scientists have asked how complete human beings could arise from fertilized eggs. The Greeks decided that each sperm had within it a tiny human, invisible to the naked eye. The Roman orator Seneca wrote: "In the seed are enclosed all the parts of the body of the man that shall be formed. The infant that is borne in his mother's wombe has the rootes of the beard and hair that he shall weare one day. In this little masse likewise are all the lineaments of the body and all that which posterity shall discover in him."

Aristotle, too, wondered "how the plant is formed out of the seed or any animal out of the semen." The question, he said, is not "out of what the parts of an animal are made, but by what agency."

But Aristotle had a different view, one that arose from his own observations as he broke open eggs at different stages of development. It made no sense, he decided, to postulate that the creature existed fully formed in the seed. After all, he wrote, "some of the parts are clearly visible as already existing in the embryo while others are not." And, he added, "that it is not because of their being too small that they are not visible is clear, for the lung is of greater size than the heart, and yet appears later than the heart in the original development. Since, then, one is earlier and another later, does the one make the other, and does the later part exist on account of the part which is next to it, or rather does the one come into being only *after* the other?"

And yet the theory of preformationism, as Seneca's idea was later called, took hold and was so powerful that even when the microscope was invented, two thousand years after the time of the ancient Greeks, the first scientists who looked at a sperm cell insisted that they saw a minuscule homunculus inside.

It was not until the waning years of the nineteenth century, a time when embryology was ascendant, that scientists began to ask probing questions about embryo development and inquiries began that led, eventually, to the idea of cloning. Some of the great thinkers of the time were awestruck by the gradual, yet perfectly orchestrated growth of a complex creature from a single fertilized egg. They were hobbled in their attempts to study embryo development because they had no way to grow mammalian embryos in the laboratory—even unfertilized egg cells would soon die when they were removed from an animal's body. But members of the cold-blooded kingdom of amphibians—frogs, toads, and salamanders—were the perfect creatures for these biologists. Their eggs are huge: A frog's egg is almost two millimeters in diameter, visible to the naked eye. A human egg, in contrast, is ten times smaller, microscopic in size, and its volume is much less than a tenth that of a frog's egg. Most important, unlike human embryos, and those of other animals that develop inside the mother's body, amphibian embryos develop in the open, so embryologists could watch the entire process. They could fertilize the frog eggs in their laboratory and then keep track as the fertilized egg began to grow, first dividing in two. Each of those cells would divide again, and thus the embryo would grow into a ball of cells. Finally, the embryo would take on a recognizable form. It would grow limbs and eyes, skin, muscles, and a spinal cord, developing, eventually, from a tadpole into a frog. Out of a single egg, this complex creature would be created

Frogs were a particular favorite because they had an additional advantage—they make enormous quantities of eggs. A single frog spews out as many as three thousand eggs in one gelatinous mass. Even today, if a cow or a mouse or a woman is given drugs to throttle her ovaries into super-ovulating, they may make ten, at most twenty, eggs in an ovulatory cycle.

Of course, frogs were not people, or sheep, for that matter. But, said Robert Gilmore McKinnell, a seventy-one-year-old embryologist at the University of Minnesota and one of the last working scientists who remembers the pre–molecular biology days, "biologists had a dogma that what you observe in one organism, if it's significant to life itself, will likely be appropriate to other organisms." And, he added, "this has proven to be the case." For example, "you don't have DNA as the genetic material of butterflies and a different material for cows."

Embryologists of the nineteenth century, focusing on amphibians, were occupied with the most basic questions. Did the cells of an embryo develop independently, each going its own way? Or did they interact, so that each cell's fate was determined by the cells around it? What, exactly, determined a cell's ultimate destiny? And how did it all fit together—how was the intricate dance of development choreographed?

In science, the hardest part of getting an answer often is deciding what questions to ask. Once a question is formulated, it can lead to a hypothesis and then to experiments designed to attempt an answer. When scientists test a hypothesis in this way to see if it is true, a new understanding of nature can emerge.

So it was in embryology. "What was so excellent about those early people was that they asked such good questions and planned their experiments so well," said Marie Antoinette Di Berardino, who is professor emerita of the Medical College of Pennsylvania–Hahneman College of Medicine.

The field became the most exciting in biology, drawing the best students and garnering the greatest interest not just among scientists but also among philosophers and even theologians. Viktor Hamburger, who was born at the turn of the century and worked with the great embryologists of his time, wrote that embryology used to reign supreme. "To biology students of my generation, it held the same fascination as molecular biology and neurobiology do today. We were impressed by the rigorous causal-analytical approach to fundamental problems of embryonic development and intrigued by the prospect of performing experiments on living embryos. We marveled at the elegance and superb craftsmanship in the performance of the masters of this art and were hardly aware of the incongruity between the enormous complexity of the developmental processes and the limitations imposed by the few techniques at their disposal."

But it was a different age, and the scientists of the time had different interests and were more prone to philosophize and to look for broad implications of their work, than scientists of today. They were professors, their salaries paid for their research, and grants were unheard of. They followed their interests, without the enormous pressures experienced by today's scientists to write grants and publish their results. "What you needed was a microscope and a little bit of support for trips to collect materials," said David Kirk, a developmental biologist at Washington University in St. Louis, who has taken a particular interest in the history of his field. One famous embryologist built a whole career "on a few pieces of glass rod and a microscope," he said.

Today, scientists are more narrowly trained and more intensely focused on the details of their research than on the larger picture. The equipment they need costs millions, their labs often have casts of dozens, they compete breathlessly to be the first to finish— and publish—what sometimes may simply be the next obvious experiment. The directors of many large labs no longer even lay hands on an actual test tube or pipette. The grinding work of doing experiments is left to the doctoral and postdoctoral students. Lab chiefs write grant proposals and give talks; they travel from meeting to meeting, expounding on the work that goes on in their lab and catching up on scientific gossip, being seen and observing the other elite lab chiefs. They work with the press offices of universities and with public relations firms whose job it is to get attention for the scientists' research. These publicists, with the encouragement of the scientists, relentlessly pursue journalists, calling them at home and at work, sending them press releases by fax, E-mail, and Federal Express, all in an attempt to draw attention to what can be the most pedestrian work. And often, when asked about the philosophical significance of their work, scientists today react as Ian Wilmut did when asked about the cloning of Dolly. All he was doing, he said, was creating a sheep that would produce a certain kind of milk.

For those who are accustomed to today's science, it is hard even to imagine that long ago time when the great embryolgists of the nineteenth and early twentieth centuries tried to solve the mysteries of life itself. The pace was stately, the grand masters stayed in their labs, and scientists often spent more time making tools than doing experiments. Viktor Hamburger, who was a student of the great German embryologist Hans Spemann, wrote that the equipment cost next to nothing, but it took endless hours to produce.

"We spent the equivalent of only a few dollars during the entire breeding season." Hamburger wrote. But, he added, "manufacturing the instruments required considerable skill, and as students we spent a good deal of time perfecting it. Spemann, who was not particularly strict in other matters, became very critical when we did not live up to his standard of perfection in this exercise."

In fact, Hamburger's mentor, Hans Spemann, whom Kirk described to me as "the most innovative and influential embryologist of the twentieth century," had to watch himself lest his fascination with the making of instruments became more important than actually doing experiments. Spemann wrote that making instruments "was not a burden but a pleasure which let me forget time, and I had to guard myself, that it did not grow beyond control and become an end in itself."

When a scientist joined a university faculty, he got a microscope, Kirk said. He also got a continuous supply of assistants, whose salaries were paid by the university. These scientists wrote enormously long papers describing their work. One of Spemann's students, Johannes Holtfreder, wrote several papers that were two hundred pages long, Kirk said. The reason for these huge tomes was that scientists were paid by the page when their papers were published. Like Charles Dickens, drawing out his serialized stories to their maximum length, the embryologists did not hesitate to expand upon and describe their work in luxurious detail.

Today, most journals pay nothing—the honor of being published is thought to be sufficient payment. Other journals charge the scientists for each published page. Scientific papers often are just five to ten pages long and the journals themselves strictly limit the length of their articles. Discussions of data often are curt and philosophy is absent. Because academic success typically depends on the number of papers a scientist publishes, many divide up their results, publishing each aspect of an experiment in a separate paper, a process scientists themselves have called "the least publishable unit." Sometimes, this process can go on for years, an abbreviated form of Dickens' serialized stories but without the narrative thread. The aim is to publish the most striking tidbits of data in the best journal possible, then to publish each subsequent piece of data that can be milked from a study in increasingly less competitive journals. Only the truly assiduous readers would notice, or read, the trail of publications emerging slowly from some large studies.

The early days of embryology were the days of the gentlemen scientists, dressed formally in coats and vests, and priding themselves on their worldly interests. The famous embryologists who studied the secrets of development were the antithesis of the Arrowsmith image, the scientist of the eponymous book, whose hero went into the woods to work alone, developing his science in complete isolation and with no thought of discussing broader philosophical questions with other intellectuals. And the scientists were the antithesis of people like the mathematician Fritz John, who once boasted that it was a badge of honor to work in a field so abstruse that few could even understand its language or its research questions. John declared that he sought neither fame nor fortune but just "the grudging admiration of a few close friends." Instead, those early embryologists sought the company and the comments of other great thinkers and purposely tried to envision their work in a philosophical context.

Science and philosophy, in fact, were intermingled to such an extent that, Kirk said, "no serious scientist would think of publishing his results without trying to draw philosophical conclusions from them." These scientists, he added, studied philosophy along with science and took the title Doctor of Philosophy very seriously. For much of the nineteenth century and in the early part of the twentieth century, science was actually called natural philosophy. And "at least until Darwin's time, a majority of biologists thought that what they were doing was elucidating God's handiwork, approaching the mind of God by studying His creations," Kirk said.

Even Spemann, whose entire body of work was reductionist in nature, searching for the simplest mechanical explanations for his results and eschewing the notion that living beings are different in kind from the nonliving, eventually revealed that he too was awestruck by the mysteries of life. He ended his magnum opus, a book published in 1938 called *Embryonic Development and Induction,* with a philosophical, almost mystical, passage, that said, in essence, that the forces active in the development of an embryo are not like those forces in physics and chemistry that we know and understand most clearly. Instead, he said, it is almost as though the cells of the embryo have psyches.

"Again and again," Spemann wrote, "terms have been used which point not to physical but to psychical analogies." And, he cautioned his readers, "This was meant to be more than a poetical metaphor. It was meant to express my opinion that, even laying aside all the philosophical conclusions, merely for the interest of exact research, we ought not to miss the chance given to us by our position between the two worlds. Here and there this intuition is dawning at present. On the way to the high new goal I hope to have made a few steps with these experiments." And so his book ends.

These embryologists were speaking to more than just their fellow scientists. They were working in the decades around the turn of the century when the scientific world was in a period of upheaval. Charles Darwin had published his theory of natural selection, psychologist William James was developing his theories of the self, and Sigmund Freud was developing his theories of the unconscious. Thinkers from these seemingly unrelated disciplines drew freely upon each other's work, using ideas from one arena to help justify ideas in another.

The first breakthrough in scientists' attempts to explain development came in the form of a theory, articulated by August Weismann, a German professor of zoology and comparative anatomy at the University of Freiberg, in the waning years of the nineteenth century. Weismann's theory was a seminal event in embryology— a theory so rich and so cleverly crafted that it all but begged for experiments to test its predictions. It was a theory so provocative that it culminated, more than fifty years later, with the first thought of cloning.

Weismann was a serious man, with a furrowed brow, wire-rimmed glasses, and a clipped white beard. A senior professor, he was accorded respect nearly to the point of obsequiousness in the German university system. His ideas about heredity and environment were fundamental; Darwin himself quoted Weismann approvingly on the first page of his book *On the Origin of Species,* stating, "As Professor Weismann has lately insisted, and as I have incidentally shown in my work," organisms are under the influence of both heredity and environment, but heredity "seems to be much more important."

Weismann was obsessed with the notion that the key to understanding development is to understand why it is that a cell of an adult—a brain cell, for example—is a brain cell forever. After all, every cell originates from a fertilized egg. What makes a cell differentiate and its fate be sealed? Why does the arrow of development point in just one direction? Why can't cells go backward as well as forward in development, with a skin cell turning back into an embryo cell, for example, and then perhaps going on to become a liver cell?

In 1885, Weismann hit upon a clever solution. The only way

that development could be so unidirectional, he reasoned, would be if the actual genetic information in a cell changes, diminishes, as a cell differentiates. He proposed that a fertilized egg contains all the information needed to produce a complete individual. But, he suggested, there must be a sort of a sequence of bisections of the egg's nucleus so that, with every cell division, the daughter cells have less genetic information than the parent.

His theory said that the diminution of genetic information occurs as soon as a fertilized egg first divides, forming a two-celled embryo, each cell of which is called a *blastomere,* from the Greek term for "part of a bud." Starting with that first cell division and continuing with each subsequent division, Weismann said, a cell's genetic material divides so that each daughter cell has less information than the parent cell. He wrote: "In each nuclear division the specific plasm of the nucleus would be divided, according to its nature, into unequal moieties, so that the cell bodies, the character of which is determined by the nucleus, would thus be new stamped."

According to Weismann's theory, when an egg divides into a two-blastomere embryo, the cell on the right side would contain all the information necessary to make the right side of an embryo and the cell on the left would contain all the information necessary to make the left side of an embryo. When those two cells divided, each of the four cells that resulted from that division would contain the information needed to make one fourth of an embryo. The process would continue, so that the more cells that were created, the less genetic information each would contain. In the end, a liver, for example, would consist of cells that contained only enough information to be liver cells; the skin would be made up of cells that had only enough information to be skin cells; the eyes could be nothing but eyes, the brain nothing but brain, and so on.

It was a hypothesis that explained the seemingly unexplainable, and it hit a chord with embryologists and even psychologists. In his book *Principles of Psychology,* William James used Weismann's ideas to attack Darwin's nemesis, Jean-Baptiste Lamarck, who said that animals, and people, could inherit acquired characteristics. Criticizing Lamarck's theories, James wrote that Weismann "has a captivating theory of descent of his own which makes him think it *a priori* impossible that any peculiarity acquired during lifetime by the parent should be transmitted to the germ."

And so, as always happens when an idea is enticing, scientists soon were devising ways to test it. What did Weismann's hypothesis predict? What sorts of experiments could determine whether his theory was true or false?

Within a few years, scientists had hit upon a contradiction. Some found confirmations of Weismann's theory; others found evidence that refuted it. Yet all were excellent scientists and their experiments seemed airtight. It was, said McKinnell, a perfect example of the fiendish difficulty of doing experiments in embryology. In fact, he wrote, "The fact that few Nobel Prizes have been conferred upon embryologists is not a result of failure to ask appropriate questions but reflects on the difficulty of obtaining answers."

The first evidence supporting Weismann's hypothesis came from an intriguing observation by a young German cytologist, Theodor Boveri, who was studying a worm, *Ascaris megalocephala,* that infects horses. When the embryo cells of that worm develop, he noticed, the chromosomes shrink—they lose genetic information as the worm's cells become more specialized. It made sense, Boveri realized. It must be that the chromosomes were shrinking because, as Weismann predicted, specialized cells have lost the genetic information they started with early in life.

The worms, Boveri discovered, start their embryonic life with two large chromosomes in their cells, one inherited from each parent. Yet by the time the worm embryo consists of just four cells, the chromosomes are changing. One cell of the embryo remains normal—it keeps its large chromosomes. But the chromosomes in the other embryo cells undergo what Boveri called "chromosome diminution," in which they break up into fragments that are divided among the cells of the developing embryo. A few cell divisions later, the already fragmented chromosomes shatter once again. By the time the embryo contains thirty-two cells, thirty of its cells have diminished chromosomes and two remain with large chromosomes. Those two become the sperm and egg cells, ready to seed a new embryo. The others become the differentiated cells of the worm's body. Boveri also noticed that he could predict what sort of adult cell an embryo cell would become by noticing when its chromosomes had shrunk.

It was seductive evidence but it was not direct proof of Weismann's hypothesis. Boveri, after all, did not prove that the cells with smaller chromosomes actually had less genetic information.

But in the spring of that same year, a German embryologist, Wilhelm Roux, had a brilliant idea that could provide the perfect evidence—or refutation—of Weismann's theory. He used the fertilized eggs from the edible frogs, known to scientists as *Rana esculenta,* to see if he could divide the embryos in two and coax them to develop independently into separate frogs. If Weismann's hypothesis was correct, it wouldn't work. The theory predicts that once the eggs divided to form two-celled embryos, each cell would have less genetic information than the original fertilized egg. Each would be missing some of the information needed to make a complete frog. Thus, if Roux broke up the embryos, he should not get new, intact frogs growing out of them.

It was mating season, and the frog eggs, lying in the nearby ponds, had been fertilized. It was time for the crucial experiment. Roux waded into a pond, collecting fertilized frog eggs. He brought them back to his lab and watched them, waiting until they had divided to form two-celled blastomeres. Then he took each blastomere and destroyed half of it, puncturing one of its cells with a hot needle. Weismann's hypothesis predicted that the cell that remained would have only enough genetic information to direct the development of half of a frog; it should develop into half an embryo.

And it did. Each severed blastomere really did develop into half an embryo. It was a moment of triumph. Weismann's theory, in Roux's experiment, appeared to be correct.

Roux and his students were ecstatic. They began repeating the experiment in endless variations. Roux also founded a scientific journal and named the new field he thought he had founded *Entwickslungmechanic,* or "developmental mechanics."

Roux's discovery rippled through the world of the intelligentsia. Even Freud appropriated his findings, using them to explain the importance of early experiences in molding the adult personality. In his *General Introduction to Psycho-Analysis,* Freud argued that "infantile experiences" should be "particularly appreciated."

Events that occur during infancy, Freud wrote, "are all the more pregnant with consequences because they occur at a time of uncompleted development, and for this very reason are likely to have a traumatic effect." He continued: "The work done by Roux and others on the mechanisms of development has shown that a needle pricked into an embryonic cell-mass undergoing division results in serious disturbances of the development; the same injury to a full-grown animal would be innocuous."

But the story soon became muddled. Another scientist, working with another species of amphibian, got results that seemed to contradict Weismann's theory. Hans Adolph Eduard Dreisch, inspired by Roux's success, was working with sea urchins. Since sea urchin eggs are much smaller than frog eggs, Dreisch did not have the option of destroying one cell of a two-celled blastomere by poking it with a hot needle. He realized, however, that he could shake the cells apart. He put the embryos in a beaker of sea water and shook it long and hard, until the embryo cells separated. Then, he asked, would the single-celled embryos develop as half sea urchins?

To Dreisch's astonishment and to the shock of the scientists worldwide, the expected did not happen. The embryos, which had been torn asunder, developed into complete, whole sea urchin embryos, albeit dwarfs that were smaller than normal.

Dreisch tried his experiment again, this time using four-celled embryos and shaking each apart to form four separate cells. Once again the individual cells grew into diminutive, but complete, embryos.

It made no sense. If Roux's frog study was correct, then these separated sea urchin embryo cells could not have enough genetic information to develop into whole embryos. If the frog experiment was not correct, then why did the frog embryos develop into half embryos and not whole ones?

Dreisch thought that perhaps the problem was that Roux's poking the frog embryos with a needle had altered the results, somehow injuring the remaining cells and preventing them from developing properly. In that case, he decided, if he separated frog embryos in two by shaking them just as he had the sea urchin embryos, maybe they would grow into complete frogs.

But Dreisch was unable to do it. The frog blastomeres simply would not break apart. "I have tried in vain to isolate amphibian blastomeres. Let those who are more skillful than I try their luck," he moaned. He wrote that not only was Weismann wrong but that Weismann's entire view of the living world was wrong. There was a life force, a vitalism, that drove development. Life was not the result of the normal laws of physics and chemistry that governed nonliving things. Development, Dreisch wrote, is attributable "not to elementary physico-chemical laws but to elementary laws of living matter, to vitalistic elementary laws." With that, he left the field of embryology entirely, deciding that the rules of life would not yield to the experiments of science. He turned in solace to philosophy.

Viktor Hamburger, who knew Dreisch as a professor at the University of Leipzig, was intrigued by Dreisch's conversion to vitalism, the belief that there is a mysterious life force that animates living beings, and by Dreisch's ultimate fate. Hamburger took Dreisch's philosophy course at the university and heard Dreisch make his vitalistic arguments, but remained unconvinced by Dreisch's reasoning. A friend of Hamburger's, who had written his doctoral thesis with Dreisch on a topic related to vitalism, was similarly unsuccessful in persuading Hamburger to reconsider Dreisch's philosophy. Hamburger wrote that his friend "knew Dreisch well and conveyed the image of a fascinating, congenial personality; he told me of Dreisch's cosmopolitan outlook, his liberal political leanings, and his dedication to pacifism that, later on, brought him on a collision course with the Nazi regime." Eventually, Dreisch lost his professorship at the University of Leipzig because of his political leanings. In his waning years, Dreisch was drawn to parapsychology and the occult and he spent the last two decades of his life in a monastery.

Yet Dreisch's sea urchin embryo experiment, which changed his life and lured him away from science and into the world of the occult, was correct. The frog embryo experiment of Roux had given misleading results.

Although Dreisch, by then, had turned his back on mechanistic biology, embryologists gradually realized the truth of his experiment as, over the next few decades, they managed to separate the cells of blastomeres of frogs and salamanders. Each time, they found that the individual cells would develop independently and turn into normal animals. And so, it turned out, Weismann was wrong—the cells of an embryo do not lose genetic information as they divide. At least in the early stages of development, each cell retains all the genetic information it needs to make a complete animal. But, of course, these studies involved only the earliest embryo cells, taken when an embryo contained two to four identical cells. These cells had not begun to specialize, so no one could say for sure whether a specialized cell from a fetus or an adult also had retained all the genetic information present in a fertilized egg.

Why had the frog embryos failed to develop in Roux's original study? In hindsight, McKinnell said, it is clear what went wrong. The hot needle had not only destroyed one cell of the two-cell embryo; it had also put a roadblock in front of the remaining cell, preventing it from growing into the space occupied by the dead cell.

"Roux was a good reporter who failed to draw the right conclusions," McKinnell wrote. "The most accurate explanation for Roux's results seems to be that the mass of the dead blastomere, in intimate contact with the living cell progeny of the surviving blastomere, physically prevented movements of cells and otherwise inhibited the viable half from fully expressing its genetic capabilities. Blastomere *separation* is a better test of the developmental capabilities of a frog-embryo fragment."

Dreisch was completely vindicated in 1902 when Hans Spemann, who had been a student of Boveri and was the only embryologist to win a Nobel Prize before 1986, finally succeeded in dividing a salamander embryo in two. Spemann had actually become an embryologist because he was fascinated by Weismann's theory, and so it was, in a sense, the perfect victory for him to finally resolve the question of whether that theory was correct.

Spemann had had tuberculosis in the winter of 1896 and 1897 and had had to go to a sanitarium. He happened to bring just one science book along—Weismann's book, published in 1892, *The Germ Plasm: A Theory of Heredity.* It won him over to the field. Spemann wrote, "I found here a theory of heredity and development elaborated with uncommon perspicacity as to its ultimate consequences."

And so, when the debate between Dreisch and Roux was in full force, Spemann was perfectly situated to resolve it. Since it was impossible to shake the salamander embryos apart, Spemann decided to slice them in two. He took a hair from the head of his newborn baby boy and tied it into a noose. He slipped it over a two-cell embryo, gradually tightening it until the embryo split in two. Each of the embryos developed into a normal salamander.

Spemann had already experimented with this method, but had stopped short of actually severing the embryos, not realizing that he held the key to resolving the paradox of Dreisch's results. Instead, Spemann was exploring the bizarre consequences of meddling with embryos. He had discovered that if he merely pinched the embryos with his noose, they sometimes grew into two-headed monsters. Hunched over his microscope, he watched the strange creatures with fascination, observing how the two heads fought each other for control. And, he reflected, the experiences of the monster salamander embryos might inform him of the very nature of the psychological state in humans known as the self. When there are two heads and one body, what becomes of the self? What, in fact, does "individuality" mean?

Spemann wrote of his experiences: "Such animals came to the stage of feeding, and it was now most remarkable to see how once the one head and at another time the other caught a small crustacean, how then the food moved through the separate foreguts to the joint posterior intestine. . . . It was probably irrelevant for the well-being of the strange double creature which head caught the food; it was of benefit to the whole. Nevertheless, one head pushed the other away with its fore legs. Hence two egotisms in the place of one." And so, he continued, "the interest was heightened by the occasional occurrence of such double monsters in man. Here too, a simple intervention would have the same disquieting consequences. Thus at last—I was then 28 years old—I had found the beginning of my own scientific journey. . . . It was first the fascination with the mystery surrounding the 'partially split individuality,' then the enjoyment of the elegant experimental technique, but then simply the continuing commitment which forced me to seclude myself in my room, one spring after another, and, instead of roaming in the lovely world, to bend over the binocular microscope and tie hairloops around the slippery eggs of salamanders, until I had constricted about a thousand and a half."

When he sliced a salamander embryo in two, Spemann had shown that early embryo cells retained all the genetic information necessary to create an entirely new being. But the next question was, What happened to genetic information in the cells of older embryos? Did they, too, retain all the information that was present when a sperm fertilized an egg?

It turned out that they did, and the evidence came from primitive cloning experiments that were accidents, occurring when a scientist was studying parthenogenesis—the creation of embryos from eggs that had not been fertilized, in effect, trying to create embryos that had only mothers, no fathers.

Working at the turn of the century in the gray stone halls of the University of Chicago, a German-born embryologist named Jacques Loeb found a way to trick unfertilized sea urchin eggs into dividing as though they had been penetrated by a sperm. He would mix the eggs in solutions of seawater and magnesium chloride, which, he found, could give the unfertilized egg a jolt and shock it into dividing and starting to develop. But this rough treatment would sometimes tear the egg's membrane, herniating the egg so that some of the cytoplasm would bulge out of it. McKinnell describes this as an egg with an "appended bleb" of cytoplasm.

The egg would start to divide, forming an embryo. The appended bleb of cytoplasm would be carried along, bulging out from the embryo. Sometimes, however, when the nucleus of one cell of the embryo divided but the cell itself had not yet cleaved in two, one of the newly formed nuclei would ooze into the bleb of cytoplasm that protruded from the egg. And sometimes, this newly nucleated cytoplasm would break off from the egg, turning into an identical-twin embryo, which would develop on its own. It was, McKinnell said, "a cloning experiment of nature" in which the nucleus of one cell was used to create an entirely new being, identical to the original. And it was further proof that Weismann's hypothesis did not hold for embryo cells. The newly created embryo had the genetic material from an early blastomere—not from a fertilized egg. Yet that genetic material contained all the information necessary to create a new sea urchin.

Spemann wondered whether higher organisms, like a salamander, which is a vertebrate, would behave the same way. Could he create a new, identical-twin embryo by blebbing out the cytoplasm of a salamander embryo and waiting for a nucleus to cross into the cytoplasmic bulge? If so, it would demonstrate, once again, that an early embryo contains all the information necessary in each of its cell's nuclei to specify the development of a new creature.

So Spemann again used a baby's hair, drawing it into a noose. He enriched a newly fertilized salamander egg, gradually tightening the noose, forming a dumbbell-shaped cell and forcing the nucleus to one side.

Then Spemann watched as the egg started to divide. Only the half that housed the nucleus divided, but it grew to a ball of sixteen cells. Then, carefully, Spemann loosened the noose so that one of the nuclei from the embryo slipped into the protruding bleb of cytoplasm. When the nucleus in the bleb began to divide, Spemann tightened the noose again, completely severing the nucleated bleb from the larger embryo. And then, he discovered, the nucleated bleb developed independently, an identical twin of the larger embryo.

Spemann had conducted a primitive cloning experiment. He had transferred a nucleus from a cell of a sixteen-cell embryo to a cell that had no nucleus. And he had demonstrated that a nucleus from a developing embryo was capable of directing the development of an independent salamander.

Spemann wished he could go further. He realized that it was not enough to show that cells from a sixteen-cell embryo retained an ability to direct development. After all, he said, it might be that cells from older embryos, whose cells had become specialized, had lost that ability. And so, ten years later, Spemann reviewed his work in his 1938 book, *Embryonic Development and Induction.* There, he first proposed cloning, of doing what he called "a fantastical experiment." It would involve removing the nucleus from a differentiated cell—a cell from a more advanced embryo that was already taking the shape of a salamander, or, even more fantastically, a cell from an adult—and placing it into an egg whose nucleus had been removed. Then, he wondered, would a normal embryo develop? Could the nucleus of a differentiated cell direct the development of a new creature?

"Decisive information about this question may perhaps be afforded by an experiment which appears, at first sight, to be somewhat fantastical," he wrote. "This experiment might possibly show that even nuclei of differentiated cells can initiate normal development in the egg protoplasms."

But Spemann could not think of how to do it. He wrote: "The first half of this experiment, to provide an isolated nucleus, might be attempted by grinding the cells between two slides, whereas for the second, the introduction of an isolated nucleus into the protoplasm of an egg devoid of a nucleus, I see no way for the moment."

It was not until 1952, eleven years after Spemann's death, that cloning—adding an isolated nucleus to an egg whose nucleus had been removed—succeeded.

Robert Briggs thought cloning was truly a long shot and so did the National Institutes of Health, which paid for his research. Briggs was an embryologist with a Harvard Ph.D., working at the Institute for Cancer Research and Lankenau Hospital Research Institute in Philadelphia in the 1950s, a time when modern science was getting under way. No longer were embryologists august professors whose research cost little and was paid for by their universities. Now scientists like Briggs were utterly dependent upon the federal government—in particular, the National Institutes of Health—for grants. Science was becoming much more aggressive and competitive. In order to get a grant, the lifeblood of any laboratory, scientists had to tell the funding agency in great detail exactly what they intended to do. If an expert panel of scientists who reviewed the proposal did not think it likely enough to succeed, the proposal was not funded and the experiment could not go forward.

Briggs wanted to understand how genes are activated and inactivated during development. How does an embryo cell "decide" to use or not use certain genes that specify the very nature of the cell and its special functions in the growing organism? He discussed the question endlessly with his colleague, Jack Schultz. Eventually, Schultz suggested that Briggs transplant a nucleus from a frog cell into a frog egg to see if the cell's DNA had irreversibly changed during development or whether a normal frog would develop from the union. This, of course, was exactly the "fantastical experiment" Spemann had dreamed of doing, but neither Briggs nor Schultz knew about that. They were working in a different era, when science had already become specialized and the philosophy-based science of the past was on its way to being forgotten. And so, Marie A. Di Berardino, a young scientist in the Briggs lab at the time, told me, they were unaware of Spemann's notions.

Briggs worked with embryos of the northern leopard frog, *Rana pipens.* This is the spotted frog found in ponds in the United States and Canada, popular with scientists because it is so easy to obtain. After working for several years with the embryos of these frogs, Briggs was ready to try to transplant a frog-cell nucleus to an egg cell. He needed a junior scientist who could do the delicate microsurgery, and he found one—Thomas J. King—who was studying for a Ph.D. at New York University and who had taken a course in the technique.

Briggs applied to the National Institutes of Health for funds.

His application ended up at the National Cancer Institute, where it was summarily dismissed, Di Berardino said. One reviewer commented that it was "a harebrained scheme with little chance of success."

Undeterred, Briggs sent in his grant application again. This time, the cancer institute sent a delegation to Briggs's lab to see if there was any way to justify funding the work. They finally acceded to Briggs's request, reasoning that after all, he was asking for very little money, just enough to support King. So, Di Berardino said, "Finally, on February 1, 1950, Tom King was appointed research fellow in the Department of Embryology," and the famous experiment began.

Briggs knew what kind of cells he wanted to work with—blastula cells. They were not cells from adult frogs because, he decided, he wanted to start with experiments that were most likely to succeed, then move on to the chancier ones. The blastula cells were from embryos containing eight thousand to sixteen thousand cells, early enough in development that the frog embryos that did not appear to have begun what might be an irreversible process of forming specialized tissues and organs. A nucleus from one of these cells, Briggs reasoned, might be able to support the development of a new frog if it was inserted into a frog egg.

The cloning experiment was conceptually easy. All Briggs would have to do was siphon the nucleus from an unfertilized frog's egg with a glass pipette, then add a nucleus from a blastula cell to the egg. He obtained a blastula cell nucleus by first breaking a frog embryo apart into single cells and then pulling one cell into a pipette that was narrower than the diameter of the cell, but not narrower than the nucleus. The cell broke apart as he sucked it into the pipette, but its nucleus remained intact.

But to do the experiment without damaging either the egg or the nucleus that was to be added to the egg would be "a technical feat," Di Berardino said. "There wasn't any background in the literature to say that this was a sure bet." And so, she said, "Briggs had a number of other projects going to make sure that things didn't fold up" if the cloning experiment failed.

Briggs and King made their tools by hand—glass needles and micropipettes. After they used the needle to suck the nucleus out of a frog egg cell, they used a fine scissors to cut the jelly layer that coats the egg. They pulled the jellylike film aside with a watchmaker's tweezers. Then they inserted a micropipette containing the blastula cell nucleus into the interior of the egg and plopped the nucleus inside.

At first, as so often happens in science, nothing worked. The frog embryos simply died after dividing a few times. But finally, one evening in November 1951, one of the eggs seemed to be surviving and developing. King was ecstatic, Di Berardino said, and "beseeching the nuclear transplant to behave properly through the night, he returned home." The next day, the embryo was still alive and growing. Crowds of scientists came into the lab to see it for themselves.

Then tragedy struck. "After the viewing crowd had gone, Tom reexamined his prize, but found to his horror that someone had crushed the very first cloned blastula with a pair of forceps. We gently patted Tom on the shoulder and said, 'If this is real, you can do it again.' " He did, a month later.

It was a time of triumph, captured in a poem that King wrote and presented to his mentor, Briggs, just before King left Philadelphia to go to Indiana University in 1956. The poem was based on "A Visit from St. Nicholas," only it involved clones, not Santa. It began

'Twas sometime before Christmas in the year '51
Nothing was working, not even arabic gum

The poem continued, describing the first cell divisions of the first successful clone, then saying,

> It was chubby and plump, a right jolly old tad,
> And we were to it—both Mother and Dad
> A twist of its tail and a turn of its head
> Soon gave us to know we had nothing to dread;
> We spoke not a word, but went straight to the car
> And drove out to the pike, to our favorite bar
> Now it isn't the best, and it isn't a Mecca
> But they make good Martinis, with M and R seca
> We toasted the eggs, and we toasted the lab,
> And we came very near to going home in a cab.
> But we knew at the time, although slightly tight,
> We'd never forget that memorable night.

Briggs and King eventually transferred blastula nuclei into 197 frog eggs, 104 of which divided and started to develop. Of them, 35 became embryos; 27 became tadpoles. Since it takes two to three years for fertile frogs to develop from tadpoles, the scientists declared the study complete when they saw tadpoles emerge from the frog embryos. The "fantastical experiment" had succeeded, albeit with embryo cells. But it would turn out to be much more difficult to clone frogs from fully differentiated cells.

King, a taciturn man who now lives in retirement in a Washington suburb, told me that he and Briggs were ecstatic when they cloned their embryo cells and did not realize at first how hard it would be to clone more developed cells.

"We got a good deal of reaction, from both scientists and nonscientists," King said. "They thought it was phenomenal. We thought that we could clone any cell," he recalled.

In the next few years, embryologists scrambled to repeat the cloning studies of Briggs and King. They tried it with a different frog species—and it worked, indicating that the phenomenon was not unique to the type of frog that was being cloned. But over and over, the scientists found that the more advanced the embryo cells from which they drew the nuclei, the less likely it was that the clones would grow and develop into frogs.

The falloff was dramatic. For example, 0 to 2 percent of the cloned cells from young tadpoles developed into tadpoles, whereas as many as 44 percent of cloned cells from early frog embryos did so. Moreover, the older and more specialized the cloned cell, the more likely it was that the frog embryos would become developmental monsters, if they grew at all, and the sooner it was that the cloned embryos simply died.

For Briggs and King, the lesson was clear: Weismann was partially correct. There must be a diminution of genetic potential as an embryo develops because the older and more specialized the cell, the harder it is to clone.

The implications were becoming obvious. Yes, early embryo cells could be cloned. The DNA of these cells could be reprogrammed. But DNA seems to pass a point of no return as an organism develops. Eventually, when a cell is differentiated, its DNA cannot turn back.

McKinnell wrote about these results in a book published in 1979: "Many scientists have speculated about why adult nuclei seem less capable than younger nuclei of promoting normal development when transplanted to egg cytoplasm," he began. It could be that the DNA itself may be "altered or rearranged" as cells develop. And so if DNA is altered in a variety of cell types as a consequence of development, he was led to a conclusion: "The cloning of adults (human and nonhuman) may be impossible for genetic reasons."

In other words, if the results of the frog cloning experiment held true in other species, it would be impossible to clone an adult mammal.

In the late 1960s, John Gurdon, a developmental biologist at Oxford University, tried the experiment again, this time using cells from the intestinal lining of tadpoles of the South African frog, *Xenopus laevis,* a favorite creature of developmental biologists because it develops quickly. These tadpoles were old enough to begin feeding, so their intestines presumably were differentiated. In 1962, in a famous experiment that was to be cited for decades by developmental biologists, Gurdon reported that he obtained fully developed, sexually mature frogs by cloning—transferring intestinal cells into frog eggs. Granted, the experiments only worked about 2 percent of the time, but Gurdon argued that he had proved his point—that fully specialized cells retained all the genetic information necessary to direct the development of an adult organism. The reason that cloning had proved so difficult, he said, was not because the cells' genes had changed during development but because the manipulation of cell nuclei—sucking them out of cells, adding them to eggs—inevitably damaged them. The problem was not with the cells' genes but with the clumsy fingers of the humans mucking about with the cells.

While Gurdon's experiment became one of those historically important studies that suddenly put cloning on the horizon, others raised serious questions about it just a few years later. A student of Briggs named Dennis Smith found that he could clone only from undifferentiated embryo cells, and he noted that 2 to 5 percent of the intestinal cells that Gurdon used for his cloning experiment were of just this type.

These undifferentiated cells actually were primordial sperm or egg cells. In most species of frog, these cells form in the lining of the stomach and move to the gonads. By definition, a sperm or egg cell must have the ability to direct the development of a frog, Smith argued. The sperm and egg cells had not begun their final process of cell division, meiosis, in which they lose half of their chromosomes, so they had the complete amount of DNA necessary to create a new organism. If those undifferentiated sperm or egg cells were the ones that Gurdon was cloning, it was no wonder that Gurdon was successful about 2 percent of the time, Smith said. Gurdon was not using mature cells, as he'd thought, but instead was using immature ones.

So Smith repeated Gurdon's experiments with a different species of frog, one in which the sperm or egg cells do not migrate through the intestine to the gonads. He reported that he was unable to clone from the intestinal cells of these frogs. Gurdon's experiment, it seemed, had been discredited.

But then Di Berardino asked whether *any* of the experiments in which nuclei were transferred from cell to cell were really the correct test to ascertain whether a mature cell's chromosomes could direct development. To prove her point, she took sperm cells from the gonads of frogs, transferred their nuclei into egg cells, and found that they did not direct development either. If the nuclear transfer test was a correct test of the completeness of a cell's genetic information, her experiment should have worked. In further tests, Di Berardino discovered that nuclei are damaged after they are transferred to an egg. The nuclear transfer test, she showed, was not a good one.

Nonetheless, noted Kirk, Gurdon's experiments had caught the public's imagination. They became ensconced as scientific proof that frogs could be cloned from the cells of adults. Even the fact that Gurdon was unable to get his frogs that were cloned from adult cells to develop beyond the tadpole stage tended to be overlooked as his experiments entered biology textbooks and generations of students were informed that Gurdon had proved that, in frogs, adult cells could be cloned.

Among embryologists, however, as the 1960s wore on, the cloning situation became increasingly confused. The most generous interpretation of the results was that adult frogs could develop from cloned embryo cells and tadpoles could develop from cloned adult frog cells. But no one could do the critical experiment that everyone was waiting for: No one could clone an adult frog from an adult frog cell.

4

IMAGINING CLONES

Some biological scientists, now wary and forewarned, are trying to consider the ethical, social, and political implications
of their research before its use makes any contemplation
merely an expiating exercise. They are even starting to ask
whether some research ought to be done at all.

> —WILLARD GAYLIN
> *The New York Times Magazine,*
> March 5, 1972

If Dolly had been born in 1957, rather than 1997, it is likely
that the world's reaction would have been radically different. Yes,
theologians and some philosophers would have argued about the
meaning of cloning an adult mammal and the prospects for cloning humans, but the organized ethics movement that drove the
1997 debate, the ethics committees of professional societies that
issued opinions, even the national commissions set up by the
United States and Europe to advise world leaders about the wisdom of attempting to clone humans would not have existed. There
may have been a public debate, but it would hardly have been so

clamorous. And it is unlikely that local and national governments would have seriously considered passing laws to prohibit humans from being cloned.

Today, when ethics and ethicists seem ever-present, it is hard to imagine that not long ago people who wanted to call themselves ethicists had a difficult time being heard. Today, when the public looks with suspicion upon scientists who want to tinker with life, it is hard to imagine that just thirty years ago, many thought that tinkering with life would be a boon and that by doing so scientists could help us wrest control of human evolution.

From the ebullient 1950s to the dark 1970s the public mood was transformed and an ethics movement was born. Every sentiment uttered with the birth of Dolly was voiced in those years, and oddly enough, those sentiments often arose as people contemplated cloning. Even though cloning was at that time a scientific impossibility, it was pondered and dreamed of, taken seriously as a future prospect by scientists and by science fiction writers alike.

The cloning debates began with the frog experiments by Robert Briggs and Thomas King and later by John Gurdon, all of whom managed to clone frogs that grew to adulthood by transferring the nuclei from tadpole cells to frog eggs whose nuclei had been removed. The next step eluded all—no one had ever succeeded in cloning an adult frog from an adult frog cell. But the studies were mesmerizing. And some of the most brilliant scientists of the century were hooked.

Among the first was the near-legendary British biologist, J.B.S. Haldane, who speculated on the results of the frog studies at a scientific symposium. His speech, published in 1963, marked him as among the first to use the word *clone* to describe what was done. Of course, the idea of cloning dated back to 1938, with Hans Spemann's proposal for a "fantastical experiment." But Spemann did not use the word *clone,* which comes from the Greek word for "twig." Even Robert Briggs and Thomas King did not use the word *cloning* in the 1950s, in their experiments with frogs. They called what they did "nuclear transplantation."

Haldane was speaking at a convocation of scientific futurists, speculating on how long people might live, how people might conquer disease, and even how behavior might be controlled. His talk was entitled "Biological Possibilities for the Human Species over the Next Ten Thousand Years." And he did not shy away from controversy in making his predictions.

The cloning of humans, Haldane said, would become possible and could be a tremendous boon, enabling humans to control their own evolution. Of course we would clone the best and the brightest, probably waiting until people were at least fifty, and had demonstrated their superiority, before cloning them. Thus we would gradually increase the number of great thinkers, great artists, great athletes, even great beauties, in the population. If we cloned people with "attested ability" we might "raise the possibilities of human achievement dramatically." Of course, not every great person's clone would be as great as the original human being. Sometimes, Haldane said, a person's greatness would turn out not to be inborn but instead "due to accident." But no matter. Taken as a group, the clones of geniuses would be better than most people, and so, as these clones were created, the average achievement of the population would increase.

Haldane even argued that the clones of great people would have an advantage growing up. "For exceptional people commonly have unhappy childhoods, as their parents, teachers, and contemporaries try to force them to conform to ordinary standards. Many are permanently deformed by the traumatic expectations of their childhoods. Probably a great mathematician, poet, or painter could most usefully spend his life from fifty-five years on in educating his or her own clonal offspring so that they avoided at least some of the frustrations of the original," he said.

A few years later, another eminent scientist took up the cloning

theme and, like Haldane, saw cloning as a way to improve humankind. First in an article in *The American Naturalist,* published
in 1966, and then, a year later, in a column he wrote for *The
Washington Post,* Nobel laureate Joshua Lederberg speculated that
the cloning of people might be on the horizon—and that it might
be a good thing for humankind. He noted that although Briggs
and King and Gurdon had cloned frogs from tadpoles, neither
group of scientists had succeeded in cloning from adult cells, but
he speculated that technical difficulties might be overcome. "Such
experiments might be made to work in man, perhaps in a few
years," he wrote. And in keeping with the general tenor of the
times, he expressed no alarm at all at the prospect, treating it
almost as a lark.

"It is an interesting exercise in social science fiction to contemplate the changes that might come about from the generation of
a few identical twins of existing personalities. Our reactions to
such a fantasy will, of course, depend on who is immortalized in
this way—but if sexual reproduction were less familiar, we might
make the same comment about that," Lederberg wrote.

Today, most scientists would scoff at the leaps of faith necessary
to contend that those experiments with frogs could possibly mean
that human cloning was a real possibility. After all, no one had
ever cloned an adult frog from adult frog cells. The frog studies
had not shown that fully differentiated cells can be made to revert
to their embryonic state. Moreover, it was entirely unclear whether
the clumsy manipulations that succeeded with the comparatively
massive frog eggs would succeed with minuscule and delicate
human eggs. And, of course, frogs were amphibians, evolutionarily
distant from humans. At that time, no one had even succeeded
in fertilizing a human egg in the laboratory, much less getting a
fertilized human egg to survive long enough to be implanted in
a woman's womb. Certainly, before thinking about manipulating
human eggs and embryos to produce clones, at the very least

scientists should be able to keep an egg alive outside the body long enough to fertilize it.

But Haldane and Lederberg were speaking in a more optimistic age that seems, in many ways, to be lost in the mists of time. When they pointed to the benefits of cloning and predicted that cloning was imminent, science and its marvels had been ascendant for two decades. Scientists, and much of the public, expected that discoveries about the natural world would inevitably improve the lot of humans. It was, perhaps, a sign of the times that a newspaper like *The Washington Post* would even consider having a regular science column, despite what Lederberg described to me as the "periodic misgivings" of the publisher, Katherine Graham. That column is long gone and has never been revived. No major newspaper today has a science columnist.

Although scientists and humanists were by no means blind to the misuses of basic research, an organized forum for arguing about the uses of science and technology did not exist. It was, instead, left to the science fiction writers to imagine what cloning might mean, and their work was known only to the fans of that genre.

The nation's love affair with science began after World War II, when, despite the shock of science's brainchild, the atom bomb, scientists came to be seen as brilliant intellectuals who would help America win the Cold War, cure diseases, and improve our lives. And science's promises had started to come true. There was the discovery of penicillin and its sudden availability after the war, with its magical abilities to cure infections. There was the polio vaccine in 1955 which eliminated the terrifying disease that crippled and killed children. There was the dawn of the space age with the launch of *Sputnik* in 1957 and the race to put a man on the moon.

The nation was amazed by the technological feats of scientists, enthralled by entertainers like Walt Disney, whose popular television show in the 1950s featured a blithely optimistic view of the wonders of science and technology and their promise to soothe the everyday worries and bothers of life. The show's Tomorrowland song, a paean to "a great, big beautiful tomorrow," was known by an entire generation. Americans read books by popular writers like Isaac Asimov who were full of boundless optimism about the ways that science would improve our world. When men finally landed on the moon in 1969, anything seemed possible.

Many scientists shared the general air of euphoria. Government money flowed freely and students poured into science, encouraged by professors who saw the coming of a golden age. The main concern was that we might be discovering so much that life would become boring, its mysteries laid bare. When so many great discoveries were already made, what was left to challenge future generations of scientists?

"Technology was going to do all our work for us, and the problem was to keep people occupied," said Daniel Brock, a philosopher and ethicist at Brown University. The public was wildly optimistic about the powers of science and technology. "The idea was that technology was going to solve all of our problems," Brock told me.

Perhaps nowhere were these thoughts so bluntly stated than in a speech in December 1970 by Bentley Glass, the outgoing president of the American Association for the Advancement of Science, which is the nation's largest professional organization of scientists. Glass's address to the group at its annual meeting in Chicago was a remarkable speech in which he expressed no compunctions about the promise of genetics to change—and improve—the human race.

According to Glass, the looming problem for humanity was a population explosion that would force people to sharply limit their family sizes. And so, he said, when parents will be able to have no more than two children, they will want to be sure that those children are perfect. Science, he said, will come to the rescue.

"No parents in that future time will have a right to burden

society with a malformed or a mentally incompetent child," Glass said. "Just as every child must have the right to full educational opportunity and a sound nutrition, so every child has the inalienable right to a sound heritage."

Glass predicted that parents will have their fetuses screened for a myriad of genetic defects, and will abort those fetuses that are imperfect or will use gene therapy to change the genes of their unborn children. He predicted that young people, at an age when their sperm and eggs would be healthiest, will store their gametes for use when they are older. He predicted that embryos that are especially desirable, because of their perfect genetic inheritance, might be frozen for use by couples who want these ideal babies, a process he called "embryo adoption." And he had no serious qualms about advocating these eugenic practices. "The Golden Age toward which we move will soon look tawdry as we no longer see endless horizons. We must, then, seek a change within man himself. As he acquires more fully the power to control his own genotype and direct the course of his own evolution, he must produce a Man who can transcend his present nature," he said.

Even Linus Pauling, the adored Nobel laureate from the California Institute of Technology, spoke unhesitatingly about using science to improve the human race. In a paper published in 1968 in the *UCLA Law Review,* Pauling proposed in all seriousness that we tattoo the foreheads of people who carried one copy of recessive, disease-causing genes so that they would not accidentally have children with someone else who carried the same gene.

Amid all this enthusiasm, there were discomfiting signs that all might not be right with the scientific world. Problems were starting to emerge in medical studies that shocked the public and that spurred some people, mostly theologians and philosophers, to try to create a new medical specialty—ethics. It was this emergence of the ethics movement that generated the Greek chorus for the cloning debates. Out of this burgeoning movement came a new field of inquiry that soon established itself as part of the scientific landscape. So when Ian Wilmut's cloning announcement came along, the ethicists were called upon to decide how—or even whether—society, and science, should proceed.

The ethics movement had its roots in real scandals and moral dilemmas of the late 1960s and early 1970s. One of the first involved questions about the way scarce medical resources were being allocated. Medical scientists had discovered how to make artificial kidneys that could keep patients alive when their own kidneys failed, giving them a reprieve from an otherwise certain death. The problem was that these dialysis machines were in desperately short supply, so not everyone could be saved. The question was, Who should live and who should die?

In Seattle, the medical community turned to a committee of volunteers to make the choices. The committee, a group of upright citizens who later became known as a "God squad," earnestly formulated rules. They gave priority to breadwinners, family men who were fine upstanding members of the community. People who did not have a job, those who seemed unstable or who lived on the margins of society, were denied the lifesaving treatment. Men were favored over women, married over single. Who was this committee, some asked, to decide whose lives should be saved? In whose eyes was a person worthier if he had a job and supported a family than if she was a poor, unwed mother, for example, or a tattered, dirty homeless person begging on the street?

In 1972, the "Tuskegee Study of Untreated Syphilis in the Negro Male," one of the greatest horrors in American medicine, was revealed to a shocked nation. It was a government study, a project of the U.S. Public Health Service, that lasted for forty years, from 1932 to 1972, ending finally when its existence was made public. Doctors followed 600 poor black men who were living in Macon, Georgia; 399 had syphilis and 201 did not have the disease. The men were given free meals, free medical exams, and burial insurance in return for participating in the study. The ostensible purpose of the study was to observe the course of the syphilis, left untreated. Yet even when penicillin, which can cure syphilis, was discovered and even when it became widely available after World War II, the doctors did not treat the men in the study. They never even told the men that they had syphilis, only that they had "bad blood."

The Tuskegee study was a sort of ethical watershed. It demonstrated the dangers of unfettered experimentation on human subjects. And it sowed such a profound distrust of medical experiments among African Americans that, to this day, surveys show that most refuse to participate in studies and are deeply suspicious of the medical establishment. The story of Tuskegee "is like Paul Bunyan," so well known that it has taken on mythic proportions, said Larry Gambrell, an African-American athletic trainer who lives in Philadelphia. Blacks, he told me, learn about the study at their mother's knee and they learn the lesson that medical scientists may be eager to exploit the defenseless. Even many blacks who, like himself, are college educated, cannot bring themselves to trust medical researchers, Gambrell said.

Around the same time that the Tuskegee study was revealed, the details of another, known as the Willowbrook study, came to light. In order to develop a vaccine for hepatitis B, doctors had purposely infected mentally retarded children at the Willowbrook State Hospital in Staten Island with the liver virus. They asked for, and received, permission from the children's parents, explaining that sooner or later every child in that institution got hepatitis anyway. The study eventually resulted in a vaccine against this incurable disease, but the damage to the public's perception of the morals of scientists was done. Despite the fact that the children's parents had agreed to allow their children to be infected, the scientists appeared to be heartless. Once again, it seemed, the weakest and most vulnerable were being used by scientists to further their careers.

And so, said Arthur Caplan, a philosopher who is now director of the Center for Bioethics at the University of Pennsylvania, "scandal was a key factor" in spawning an ethics movement of professional people who would take on controversial topics, ask what the moral response should be, and establish procedures to prevent unethical research programs from ever getting under way. The demand was mostly for rules to govern research and for protection of human rights. But some had bigger plans in mind.

Willard Gaylin, a charismatic psychiatrist, and his colleague Daniel Callahan, a philosopher with a doctoral degree from Harvard University, decided that they needed to start an actual ethics institute that would ask more philosophical, more theological questions about where science and medicine were headed. They put together a group of like-minded people, mainly doctors, lawyers, and theologians, and formed The Institute of Society, Ethics, and the Life Sciences. Their fledgling ethics center was in Hastings-on-Hudson, a quaint town on the Hudson River, a short train ride from Manhattan. Later, the center changed its name to the Hastings Center.

But even though the public was stirred by the scandals and controversies that were arising in medical research, Gaylin and Callahan had the feeling they were swimming against the stream. Ethics was not quite marginal then, but it was "on the fringe," said Caplan. "A lot of people would say it was a fad. It was way outside the mainstream of academic respectability and it had no cultural respectability."

Major medical journals that today include articles, editorials, and letters to the editor about ethics nearly every week virtually never published anything on ethics in the early 1970s. Medical schools and philosophy departments at universities, which today include ethics as an essential part of their curriculum, seldom acknowledged its existence. Foundations did not give grants for studies on ethics. And the Hastings Center, Caplan recalled, "had no budget, no endowment. It literally existed hand-to-mouth. They were begging, asking for gifts."

What was needed, Gaylin decided, was a shock to the system, a wake-up call that involved the most dramatic, the most chilling possibility he could think of. His choice was cloning.

Gaylin's decision to seize upon cloning as an ethics issue, despite the scanty scientific evidence that it could ever succeed, would seem inexplicable today when the public mood has changed so dramatically that the tenor of those times is all but forgotten, even by many who lived through them. Yes, people were shocked when they heard about the God squad in Seattle, about Tuskegee, about Willowbrook. But their shock did not assuage a general feeling that science was good, that doctors were to be trusted, that medical paternalism was appropriate.

Now, as the century draws to an end, the public's mood and attitudes have completely changed. Some have said that we agonize about new developments to the point of paralysis. Should we test people for genes, like those that predispose to Alzheimer's disease, if we can offer no way of preventing or curing the disease? Should parents be allowed to test their children for genes that predispose to cancer? How can we protect the privacy of medical records in this age of increasing use of computers and webs of managed-care companies? Should we allow doctors to prescribe lethal drugs to dying patients so they can kill themselves?

Today, the public is both drawn to, and often frightened by, the new powers of medicine and technology. The public views scientists as isolated technocrats who rather highhandedly attempt to argue from positions of authority. Medicine is seen by many as cold and impersonal, and doctors are perceived as being more concerned with technological fixes than with the people they are fixing. This is an era that many have described as anti-science, a time when people are turning to alternative medicine, seeking unproven methods that are highly promoted by those selling them as somehow "natural" or "nontoxic," and that are advertised as being time-tested remedies that are the very antithesis of the treatments of modern medicine, which are said to arise out of sterile laboratories.

Scientists wring their hands in dismay, but few heed them. Carl Sagan, the Cornell University astronomer who was one of the great popularizers of science, wrote: "It's a foreboding I have— maybe ill placed—of an America in my children's generation, or my grandchildren's generation . . . when, clutching our crystals and religiously consulting our horoscopes, our critical faculties in steep decline, unable to distinguish between what's true and what feels good, we slide, almost without noticing, into superstition and darkness."

This is a time when, increasingly, the public has stopped looking to the future with anticipation, expecting technological miracles, and has begun instead to look back at the past with a sense of nostalgia. Karal Ann Marling, a professor of art history and American studies at the University of Minnesota, summed up the attitudes of many: "We went to the Moon and all we got out of it was Teflon pans. Our goals as a people are not these pie-in-the-sky objectives that people grew up with in the 50's. [We] settle now for a house in the suburbs and to hell with the Moon. What's the point of building a monorail if we can hardly get the car to work?"

Futurist Alvin Toffler said, "A lot of perfectly fine and decent and human people now think that technology is a negative."

But Gaylin was attempting to get a hearing in a very different age. It was a time of great optimism about the powers of science to transform the world for the better, a culmination of two decades during which science and its marvels had been ascendant. In a sense, cloning was a metaphor for Gaylin. It could illustrate the lure of science and raise questions about whether there was some research in biology whose consequences could be so terrible that it ought not be done.

In fact, it was not unheard of to use cloning as an early warning sign for a society that, Gaylin thought, might be too complacently accepting of the wonders of science. And although it might seem presumptuous for Gaylin, a psychiatrist, to pontificate about cloning, he was, in fact, picking up themes that had been enunciated just a year earlier by Nobel laureate James D. Watson, the codiscoverer of the structure of DNA. In 1971, Watson testified before Congress on the future of biology and, in particular, his contention that scientists would soon be able to clone human beings. He conceded that most scientists were not talking seriously about cloning or what it might mean. Those few discussions that had taken place, he said, "have been so vague and devoid of meaningful time estimates as to be soporific." The reason was not that scientists feared that the public would lash out against them if it thought they were about to clone humans, nor that "scientists live such an ivory tower existence that they are capable of rational thinking only about pure science." Instead, he said, it was because scientists had not fully appreciated what it means to be able to clone a frog or toad and what it means to be able to fertilize human eggs in the laboratory.

Perhaps not surprisingly, it was further scientific advances that brought home these larger implications. By the time Watson testified, British scientists Patrick Steptoe and Robert Edwards were working with infertile women, developing the treatment that we now call in vitro fertilization, or IVF. The two researchers had succeeded in fertilizing women's eggs and growing a few of the fertilized eggs to embryos that might be transferred to the women's wombs. They expected to have the first successful in vitro fertilization pregnancy that year. (Actually, the first IVF baby, Louise Brown, was not born until 1978.)

These techniques developed by Steptoe and Edwards meant that, for the first time, scientists could keep tiny, finicky human eggs alive in the laboratory and manipulate them. It meant, Watson said, that the stage was set to try the same cloning methods with humans that were done with frogs and toads. "The situation would then be ripe for extensive efforts, either legal or illegal, for human cloning," Watson told Congress. Although many might recoil from the very thought of cloning a human, there would always be some who would want to try. "Some people may very sincerely believe the world desperately needs many copies of the really exceptional people if we are to fight our way out of the ever-increasing computer-mediated complexity that makes our individual brains so frequently inadequate," he said.

So Watson urged Congress to think about what was happening and to keep the public informed as well. Whether to go forward "is a decision not for scientists at all. It is a decision of the general public—do you want this or not?" And, he warned, "if we do not think about the matter now, the possibility of our having a free choice will one day be gone."

Shortly afterward, Watson wrote an article for the *Atlantic* magazine, called "Moving Toward Clonal Man," that made exactly the same points.

But Watson's sounding of the alarms fell on almost deaf ears. Certainly, he did not mobilize the citizenry to start a cloning debate. And the letters published in response to his article were surprisingly mild, commenting, for the most part, on the role of environment as opposed to heredity in making us what we are and failing to remark on the great ethical or theological questions that cloning brings forth. One writer asked, "Would the second or third James Watson try as hard if he knew in advance he would make it?"

Gaylin, however, still thought cloning could be an issue that could put the Hastings Center on the map. Through his contacts in New York literary and journalistic circles, Gaylin managed to meet an editor of *The New York Times Magazine* and convince him to allow him to write about cloning.

Gaylin brought a humanist's perspective to the subject. Cloning, he recognized, touched some of our deepest fears, fears that had been voiced in ancient myths, in the Bible, in fairy tales, in literature. The horror that many feel at the thought of cloning a human is intimately tied to our dread of the sin of pride and of the sin of vanity.

Ever since the ancient Greeks told the tale of Prometheus, who stole fire from the Gods, people have feared the consequences of overweening pride. How much power can humans grasp, how much knowledge can they have, before they have gone too far? The story of Genesis, from the Bible, tells how Adam and Eve were expelled from the Garden of Eden for succumbing to the temptation to know too much.

The seventeenth-century poet John Milton used the theme in his epic poem *Paradise Lost,* emphasizing the power of science in giving knowledge:

> O Sacred, Wise and Wisdom-giving Plant,
> Mother of Science, Now I feel thy Power
> Within me cleere, not onely to discerne
> Things in thir Causes, but to trace the wayes
> of highest Agents, deemd however wise.

The theme occurs and recurs. The fairy tales by the brothers Grimm, dating back to medieval Germany, include a story about a fisherman who heeds a flounder's plea that he be thrown back into the sea. In return, the flounder grants the man's wishes. His wife makes the man wish first for a bigger house. Then, as her ambitions grow, she seeks political power as well as wealth. Finally, she makes her husband ask the fish to make her lord of the universe. And, at that, she and her husband are cast back into their miserable hovel.

Added to the sin of pride and the temptation of humans to become God-like is the fear of the mad scientist. Mary Shelley's *Frankenstein,* written in 1818, was a paean to the theme. H. G. Wells's tale of *The Island of Doctor Moreau* is about a scientist who glories in creating monsters. "To this day I have never troubled about the ethics of the matter," the doctor says. "The study of Nature makes a man at last as remorseless as Nature. I have gone on, not heeding anything but the question I was pursuing. . . ."

Finally, there is the sin of vanity. It dates back in recorded history at least to the myth of Narcissus, who fell in love with his own reflection. It remains a powerful theme, occurring again in Oscar Wilde's *The Picture of Dorian Gray*, about a man who sold his soul so he would not have to age. Instead, his portrait would show the marks of the years on his face.

Intermingled with these themes is the eerie noton of the doppelgänger, the mysterious double, a person who may look exactly like you but who is a stranger, doing deeds you would never have contemplated.

Cloning touched on every one of these dark themes. It would be the ultimate act of pride—of playing God—to re-create yourself. It would be the ultimate act of a mad scientist to be so oblivious to the consequences as to actually try such a thing. It would be the ultimate act of vanity to make a carbon copy of yourself. And would you have, perhaps, created your own doppelgänger?

Gaylin was more of a poet—and a prophet—than Watson, and so his article for *The New York Times Magazine* was more deeply chilling and much less matter-of-fact than Watson's popular article that had appeared in the *Atlantic* the previous year. Called "The Frankenstein Myth Becomes a Reality—We Have the Awful Knowledge to Make Exact Copies of Human Beings," it appeared on March 5, 1972, almost twenty-five years to the day before Dolly was announced to the world.

In his article, Gaylin began by discussing what he called "the Frankenstein myth." It is "the image of the frightened scientist, guilt-ridden over his own creation," he explained. Young Mary Shelley, a teenage bride, wrote her haunting story in 1818, when no scientists seriously thought that human beings could create human life. Her story was a fairy tale that warned against the universal human thirst to control nature. And yet, Gaylin wrote, the Frankenstein myth "has a viability that transcends its original intentions and a relevance beyond its original time." It raises the question, he said, of whether some research ought to be done at all. And so, he wrote: "With the serious introduction of questions of 'ought,' ethics has been introduced—and is beginning to shake some of the traditional illusions of a 'science above morality,' or of a 'value-free science.' "

Cloning was on the horizon, Gaylin was convinced, and it was none too soon for society to ask what we wanted to do about it and whether, in fact, it was a type of research that we wanted to do at all.

Gaylin cited two experiments that led him to his conclusion that the cloning of humans would soon be possible. One was conducted by a professor F. E. Steward, a cell biologist at Cornell University. Steward studied plants, and in the 1960s, he discovered that he could literally shake apart the cells of a carrot root. Ordinarily, if a seed is shaken, it will begin to develop, but no one expected that a differentiated cell would do anything at all. What Steward found, however, was that these root cells also began to divide and grow into clumps of cells, even differentiating. Finally, Gaylin wrote, Steward "succeeded in carrying one individual cell to the ultimate stage of a full-grown carrot plant—roots, stalk, leaves, flowers, seeds, and all." And so, he added, the lesson is that "any cell can, conceivably, be thus forced, once the technology is understood, to grow into a full plant." Moreover, he wrote, "what is possible with a vegetable cell is, at least theoretically, just as possible with an animal cell."

Gaylin explained that scientists already grew animal cells in the laboratory, keeping them alive and thriving on the surfaces of glass plates, as "sheets of identical cell types." He acknowledged that "the concept of growing a whole organism from one cell asexually in a laboratory would seem impossible. But that Cornell carrot confronts our incredulity. To a scientific mind, the leap from single cell to cloned carrot is greater than the leap from cloned carrot to cloned man."

The second experiment that leads directly to the cloning of humans, Gaylin asserted, was the one by John Gurdon. He described Gurdon's experiments with the cloning of cells from the intestinal lining of frogs, in which Gurdon destroyed the nucleus of an egg cell and added the nucleus of this adult, presumably differentiated, intestine cell. "Some of the cells, on division, formed perfectly normal tadpoles, some of which, indeed, became perfectly normal frogs genetically identical to the frog that donated the nucleus," Gaylin wrote, ignorant of the internal dispute in the embryology community that accompanied Gurdon's experiments. (As noted in Chapter 3, some embryologists suspected that Gurdon's cells from the frog intestines were actually primordial sperm or egg cells that move, in the species of frogs that he studied, from the intestine to the gonads.)

Yet Gaylin concluded, like Watson before him, that the possibility of cloning humans "now sounds like a rational prediction." He cautioned, as geneticists would caution after Dolly was born, that even though the genes of a clone would be the same as those of the adult that provided the cell for cloning, the clone might not be at all like the original. After all, he said, there are powerful environmental influences on development.

"The idea of seeing 'yourself' born—as has been suggested—is a joke," Gaylin wrote. "Life experience pounds, pulls, and shapes the same genetic clay into wondrous and ludicrous variations," he explained. "A genetic St. Francis clone could evolve into a tyrant. Or, more optimistically, a Hitler clone has the potential for sainthood."

Nonetheless, cloning remained "a sideshow," Callahan told me, albeit "a striking sideshow." The mention of cloning automatically tugged at people's attention, evoking a mixture of interest and repugnance. "I don't think anyone thought it would be a central issue," Callahan said. "But it was a dramatic one." And it served his and Gaylin's purposes well. "Our problem in the early days was getting people to be interested in problems they had never heard of. I think cloning was one of those issues that helped bring us to public attention."

But, Gaylin said, although he meant to provoke and to warn while simultaneously publicizing the Hastings Center, his theorizing was met with stony coldness by many scientists. Even those who did not dismiss human cloning out of hand as being scientifically impossible decided that it would be so inane to clone humans that we need not fear that it would come to pass. There would be no reason to try such an experiment, they argued.

Robert McKinnell, the embryology pioneer at the University of Minnesota who had spent his life studying nuclear transfers in frogs, wrote: "I believe it is rational to expect cancer control in the future; I believe it is sensible to anticipate new insights into the immune process; I believe that graceful aging is a reasonable hope for a greater proportion of our population in the future." But, he added, "I never expect to witness the construction of carbon-copy humans. I do not believe that nuclear transplantation for the purposes of producing human beings will ever routinely occur."

The reasons, McKinnell said, are that the human race thrives on genetic diversity. The terrible genetic sameness of clones would destroy that which makes us powerful and adaptable. And, he noted, environment plays too large a role in shaping who we become. A clone might therefore be nothing like a carbon copy of a living adult, making it questionable why, even if we could do it, anyone would want to clone humans.

Lewis Thomas, the doctor whose evocative writings appeared regularly in the prestigious *New England Journal of Medicine,* wrote that cloning had such an edge of horror that thoughts of it should be abandoned. He diffused all thought of cloning by making it almost a joke. "It is almost no comfort to know that one's cloned, identical surrogate lives on, especially when the living will very probably involve edging one's real, now aging, self off to the side, sooner or later. It is hard to imagine anything like filial affection or respect for a single, unmated, nucleus—harder still to think of one's new, self-generated self as anything but an absolute, desolate orphan. Not to mention the complex interpersonal relationship involved in raising one's self from infancy, teaching the language, enforcing discipline, installing good manners and the like. How would you feel if you became an incorrigible juvenile delinquent, by proxy, at the age of 55?"

Then there were those who conceded that perhaps, sometime in the distant future, it might be possible to clone a human, but who said they could not imagine why scientists would go along this path. There were too few scientific incentives, they said. Cloning was more like a curiosity show than real, serious science that asked questions about the nature of genes and DNA. There would be no reason to pursue it.

And maybe it was not even possible to clone. Bernard Davis, an august professor at Harvard University, wrote that the idea of cloning humans "has understandably caused a great deal of apprehension." And yet, "that possibility remains far too remote to warrant concern today." In fact, he said, returning to the ideas reminiscent of those of early embryologists, like the great German scientist August Weismann, there is some reason to believe that specialized cells have irreversibly lost their ability to regress to their embryonic state and that, as cells specialize, they undergo what he called "irreversible small duplications or losses in the content of the genome." And so, he concluded, "Cloning from adult mammals may therefore remain science fiction."

Soon only a few flickers of interest in cloning remained and only a few scientists were so moved by the process that they gave it more than a moment's thought. One was Leon Kass, a biochemist at the National Institutes of Health, who actually stopped doing biology and became a philosopher and ethicist because of cloning.

Kass's transformation began on September 30, 1967, when he was reading his *Washington Post*. He saw Joshua Lederberg's column on cloning and was horrified by what he saw as a lack of reflection about what cloning really meant. He fired off a reply, taking "strong exception to the casual and cavalier tone" of Lederberg's piece, writing, "It is unfortunate that Dr. Lederberg is either unaware [of] or unwilling to discuss the moral and political problems involved; it is shocking that he chooses to speak as if these questions are trivial, and as if they are reducible to our prejudices concerning the people who might be asexually propagated."

Because of Kass's letter to the editor of *The Washington Post*, theologian Paul Ramsey of Princeton University asked Kass to discuss cloning and other ethical questions with him. Soon Kass found himself spending more and more time with theologians and philosophers and less and less time with laboratory scientists.

"This exchange on cloning and its sequelae actually changed my career," Kass told me. When the Hastings Center was founded in 1969, he joined it. And the first report that he worked on for the Hastings Center involved cloning. It never found a publisher, however. In fact, Kass said, the entire volume of Hastings Center essays, called "Freedom and Coercion in Assisted Reproduction," "never saw the light of day."

Ramsey, too, was provoked by the cloning discussions. In June 1972, he wrote two articles for the *Journal of the American Medical Association* on the ethics of in vitro fertilization The articles, called "Shall We 'Reproduce'?" opposed in vitro fertilization and also touched on cloning as the beginning of a journey down what he saw as a slippery slope. He warned that "human procreation has already been replaced by the idea of 'manufacturing' our progeny. Unless and until *that* concept is reversed, mankind's movement toward Aldous Huxley's Hatcheries must surely prove irreversible."

An accompanying editorial in the June 5, 1972, issue noted that we have already started down the path toward cloning a human. But, it said, should we have begun? "Given the intricacies of the human mind, we doubtless possess the potential for reproducing someday, to exact specifications, a human person. But should we? Should we have taken the first step?"

One of the final papers of this flurry of interest in cloning appeared in 1974 in the British journal *Nature*. It was by Gunther Stendt, a molecular biologist at the University of California at Berkeley, who was known for his provocative views. Stendt was preoccupied by the question of what it is that so bothers us about cloning. In his paper, called "Molecular Biology and Metaphysics," he described cloning as "a fantastic facet of genetic engineering which, though it seems taken straight from the pages of Aldous Huxley's *Brave New World,* is actually likely to become a practical reality before long." And so, he said, cloning could drastically change the human population by allowing us to "abandon the old-fashioned genetic roulette of sexual reproduction" and replace the variegated human race with "identical replicas of carefully chosen, ideal human genotypes." Yet, Stendt said, he knew of no one who actually advocated cloning human populations. And why is that? he asked.

It must be, he said, that we find cloning "morally and aesthetically completely unacceptable." Perhaps at the heart of our repugnance is a curious paradox, he proposed, asking why it is that although "it would be fun to have Kant, Beethoven, Isadora Duncan, Einstein, Clark Gable, and Marilyn Monroe living on your block, the idea of having hundreds or thousands of their replicas in town is a nightmare?" The reason, he explained, "is the generally shared belief in the uniqueness of the soul. Even though the soul is incorporeal, it is supposed to fit the body, hence it is not conceivable that a unique soul should inhabit each of thousands of identical bodies."

And yet, Stendt continued, our abhorrence of cloning may not be quite so well grounded as it seems. "To oppose human cloning, however, is to betray the Western dream of the City of God. All utopian visionaries, from Thomas More to Karl Marx, think of their perfect societies as being populated not by men but by angels that embody all of the best and none of the worst human attributes." Of course, Stendt said, with sexual reproduction, there was no chance that such a city of angels could arise. With cloning, it might be possible to create such a society, yet all of a sudden it seems that that is not what we want at all. "What is wanted is the impossible, a perfect society made up of a motley of imperfect, unique souls, warts and all."

With that last sally by Gunther Stendt, serious discussion of cloning guttered out. It had been twenty years since Briggs and King cloned a tadpole and no one was near cloning even a mouse. In vitro fertilization still had not succeeded, and even if it had, why would anyone assume that just because scientists could fertilize a human egg in the laboratory they could also clone?

But there was to be just a brief hiatus before cloning reemerged, in a different context, and at a time when actual events in the annals of science encouraged the public to fear science and to mistrust scientists. More powerfully than ever before, cloning became a metaphor for unbridled science. And this time, many scientists found themselves staking out positions that would later come to haunt them.

5

THE SULLYING OF SCIENCE

Our actual experience so far does not suggest that cloning
with adult nuclei is possible, and, as has already been pointed
out, many thousands of efforts have already been made in
amphibians to test this possibility, and they have all failed.

—CLEMENT MARKERT
Yale University,
1978

On the day before April Fools' Day in 1978, a reputable New
York publisher, J. B. Lippincott, rushed into print with a remark-
able book by a thirty-four-year-old freelance writer. In his book,
David Rorvik revealed that he had participated in the unthinkable.
He had helped an eccentric multimillionaire clone himself.

The book was a secret project, but its imminent publication
had come to light nearly a month earlier. On March 3, the *New
York Post* blared in a huge headline: BABY BORN WITHOUT MOTHER:
HE'S THE FIRST HUMAN CLONE. Afterward, journalists called scientists, scientists called each other, and the public, whipped up by
stories like the one in the *Post,* waited with anxious anticipation
for Rorvik's book. The book, called *In His Image: The Cloning of*

a Man, would presumably reveal enough about the project for already highly suspicious scientists to decide whether the whole thing was a hoax or whether the almost inconceivable was true and the first human clone had, in fact, been born.

Rorvik was a respectable, credentialed science writer, but he was not well known outside of a small circle of writers and editors who covered reproductive medicine. He had grown up in Montana and graduated from the University of Montana in 1966. He studied journalism at one of the nation's premier schools, Columbia University's School of Journalism, where he got a master's degree. He worked for *Time* magazine from 1967 to 1969 before moving back to Montana and becoming a freelance writer. His work included *Brave New Baby,* published in 1971, in which he warned about genetic engineering, and, that same year, *Your Baby's Sex: Now You Can Choose,* which he wrote with a doctor, Landrum B. Shettles. It described a method developed by Shettles that, Shettles claimed, could sort sperm so that those sperm carrying an X-chromosome, which would produce a girl, could be separated from those carrying a Y-chromosome, which would produce a boy. (The method is now discredited.) Rorvik also had written an article for *The New York Times Magazine* called "The Embryo Sweepstakes," discussing the dawn of a new era of assisted reproduction that would rely on the successful births of babies by means of in vitro fertilization and genetic manipulation of embryos, neither of which had yet occurred but both of which, some scientists cautioned, were imminent.

And thus was ushered in a new wave of theological and ethical speculation about cloning. But the cloning debate at the end of the 1970s was different in tone from the earlier discussions. The mood of the public had changed, and science was becoming a fearful thing. Molecular genetics was looming as a threat to the very existence of species, as scientists learned to snip genes out of cells and add them to other cells. By coincidence, the same year that Rorvik's book was published, the world's first test-tube baby was born in England, setting the stage for a new round of debate about the awesome power of geneticists to manipulate genes and about the emerging power of biologists to shape human embryos.

At the same time, science fiction writers, picking up on remarks by scientists themselves in the 1960s, were publishing books that raised the most chilling prospects of cloning. Naomi Mitchison, the sister of biologist J.B.S. Haldane, the British biologist who had introduced the term "clone" into common parlance, wrote a book, *Solution Three,* asking what would happen if we tried to clone humans in order to control our own evolution, and Ira Levin wrote his chilling book *The Boys from Brazil,* a bestseller that postulated clones of Hitler.

So when Rorvik announced that a person had actually been cloned, the stage was set for a standoff. It seemed that people's worst fears about science and scientists had come true. Scientists were not to be trusted. If a technology was tempting enough, if it was exciting enough, someone would try to use it. To defend themselves and their profession, scientists found themselves compelled to comment, and as forcefully as possible.

When we look back on Dolly, and on why so many scientists were so surprised by Wilmut's feat, the debate of the late 1970s has to be part of the answer. In order to placate a skeptical public, which was coming to see scientists as intrinsically untrustworthy, driven more by curiosity and a perverse glee in manipulating life than by moral qualms about what is proper and appropriate, some leading scientists found themselves boxed into a corner. They had to argue publicly, even testify before Congress, that not only was it untrue that cloning had been done, but that it was ridiculous to expect that it could be done—then or probably ever. And they had to convince themselves that even if cloning could be done, they had no interest whatsoever in doing it.

It is perhaps a testament to the public's willingess to believe in the intrinsic evil of science that Rorvik's book ever gained a credulous audience. For he told a fantastic story. Rorvik claimed that one day in September 1973, a mysterious stranger called him at his home in a cabin on Flathead Lake in western Montana. The man refused to give his name but told Rorvik that he was "a fan." When Rorvik asked him why he was calling, the man said "he was getting up in years, was still single, and lacked an heir." He explained that he wanted Rorvik to find a doctor who would clone him, and that he was willing to spend a million dollars or more on the attempt. He had come to Rorvik because "he felt that, given my interest in this area and given the things I had written, I should be in an excellent position to know who the best candidates would be."

Rorvik agonized over the proposal. "As the initial shock of his proposal began to wear off, two questions loomed large: *Could* this be done, and, more important, *should* it be done? If the answer to both questions was, 'Yes,' 'Probably,' or even, 'Under some circumstances,' then I would have to ask myself: Should *I* become involved? If not, I had to wonder if it would be fitting for me to try to prevent others from becoming involved."

The man had suggested that, if nothing else, Rorvik might want to get involved for "the excitement" of the secret cloning attempt. But Rorvik wondered if he wanted to so perturb the world. He mentioned his doubts about how the project might affect his reputation as a journalist. If he agreed to help the man, he would have to agree never to reveal the name of the millionaire or the name of the scientists who cloned him. "I knew that, if my part in this came to light, I would be accused of undermining the traditional scientific ethics of full disclosure I held dear. That would be bad, the more so since I would be guilty. On the other hand, if I sought to disclose some of the details but was compromised by the need to protect and conceal my sources—out of respect for journalistic ethics I held equally sacred—I might be doubted, disbelieved, even decried as a fraud. That might be worse."

Needless to say, Rorvik overcame his qualms and agreed to help his mysterious caller. But, he wrote, he managed to win what sounds like an astonishing degree of control over the project. He said he made the millionaire, whom he called Max, agree to allow him to be a sort of ethical arbitrator. If Rorvik decided that there were serious ethical abuses, he could call a halt to the experiment. And if Rorvik insisted that the cloning experiment stop, Max had to agree not to restart it. If Max *did* clone himself and then decided he wanted to make other clones of himself, he must first obtain permission from Rorvik and the doctors who would do the work. Max also had to let Rorvik visit the clone "at regular intervals, for the purpose of reassuring myself, if no one else, that everything had worked out all right."

In retrospect, these conditions alone look a bit suspect. Why would a rich man, who could buy the services of any scientist he pleased, need a science writer who had never even studied science and who had been out of college for only seven years to put him in touch with doctors who would clone him? And even if such a man did call Rorvik, why would he agree to Rorvik's conditions? But Rorvik insisted his tale was true.

It became more and more fantastic. First, Rorvik wrote, he consulted scientists to see if cloning could succeed. He was reassured that it could, for reasons that sound familiar—they were the exact reasons put forth by James Watson, the Nobel laureate, in 1971, and again by Willard Gaylin, the ethicist, in 1972. There was the exaggerated significance of the cloning of frogs, in which, once again, the essential problem was sloughed over—that researchers could clone a tadpole from a specialized cell from an adult amphibian's cells or they could clone an adult from an unspecialized amphibian cell. But they could not clone an adult frog from an adult frog's cells. Yet, despite this, the cloning of a normal human being from cells of an adult was the point of the cloning experiment that Max had in mind.

And Rorvik cited the incipient studies of in vitro fertilization, touting the research that should lead, any day, to the birth of the world's first test-tube baby. Human eggs had been fertilized in the laboratory in 1969 but the doctors doing the work had not yet created a pregnancy that was carried to term. (In fact, the world's first in vitro fertilization baby was born at about the time Rorvik's book was published—but not when Max was having himself cloned. Moreover, for years after Louise Brown's birth, doctors said that an ability to successfully fertilize eggs in the laboratory and then produce viable pregnancies with them was a sort of black magic. There was no clear recipe and many who attempted it failed.) But in Rorvik's book, this nascent ability to fertilize human eggs in the laboratory was supposed to mean that it would be straightforward to exchange an egg's nucleus with that of a cell from an adult—in short, to clone a man in the same way that scientists had attempted to clone frogs.

Rorvik wrote that he found a scientist who was supposedly willing and able to clone Max. The man, whom Rorvik called by the pseudonym "Darwin," could have been any of hundreds of scientists. He spoke English, was middle-aged and overweight, and was respected by his peers. Darwin agreed to do the work because Max, apparently, was going to pay him handsomely. But he was uncomfortable, Rorvik reported, and "seemed torn throughout this effort between the attainment of whatever reward in all this might be and the ever-present, though variably pronounced, conviction that he was about to be, or already had been, shorn of his scientific virginity."

Of course, the experiment was supposed to be secret, so Max took his entourage to an unnamed tropical land somewhere "beyond Hawaii." Max was the master of this domain; he owned the rubber plantations there and even had an interest in the fishing industry. He had built a hospital for the native people, earning their unending gratitude. It was in this hospital that the medical experimentation was to take place.

A doctor called Mary ran the gynecological service and conveniently provided Darwin with egg donors, who were young women seeking tubal ligations after having completed their families. Unbeknownst to them, they were injected with a fertility drug to make them super-ovulate and then, while the doctors tied their tubes, they removed their ripened eggs.

At the same time, Max was looking for the perfect woman to bear his clone. Rorvik's tale, at this point, had tinges of the creepily erotic. He wrote that Max's factotum, Roberto, was in charge of finding young and beautiful girls who might be surrogate mothers for the baby. Max insisted on young virgins because he thought he might want to take the mother of his clone for a mistress, after the cloning was done. Roberto, Rorvik wrote, "would go through the factories and invite various girls to come to the clinic for examination for possible candidates in a 'study.' So many, it seemed, succumbed to his blandishments that, at one point, Darwin said, he was flooding the place with beautiful girls." Finally, Roberto found her, a young girl, "going on seventeen" who was an orphan and "very mature for her years." Rorvik called her "Sparrow."

Eventually, the cloning experiment took place, the virgin Sparrow became pregnant with Max's clone, and toward the end of her pregnancy, she and Max flew to the United States for her baby to be born. She had the baby "in a small hospital." Max tape-recorded the clone's first cries but, at Sparrow's request, did not photograph the birth. The fateful day was two weeks before Christmas, 1976. It was still two years before in vitro fertilization was to succeed, through doctors who had spent a decade trying, and it was twenty years since scientists had been stymied in their attempts to clone a frog.

Not surprisingly, scientists were outraged by the claims that the book was true. Even before they saw the book, when the only information they had came from sources like the *New York Post's* page 1 story claiming that an eccentric millionaire had cloned himself, scientists were circling the wagons, refuting out of hand Rorvik's sensational claims.

Rorvik went into hiding. His publisher issued a brief statement: "To protect the child from harmful publicity, and other participants from certain controversy, Rorvik refuses to divulge names or places even to his publisher. The story he tells is an astonishing one. David Rorvik assures Lippincott that it is true. Lippincott does not know."

Other publishers revealed that they had turned Rorvik down when he offered them his book, but they stressed that they had no certain knowledge that his book was a hoax. Jonathan Segal, then a senior editor at Simon & Schuster, said the publisher's editorial board declined the book because "we asked for verification and we weren't given it." Nonetheless, Segal said, Rorvik was a serious writer.

An article in the March 24 issue of the journal *Science,* published a full two weeks before the book's publication date (*Science* dated issues of its weekly magazine a week after they were published, so the March 24 issue was actually published on March 17) laid out some of the reasons for scientists' skepticism. Some were almost ludicrous, unintentionally revealing the egotism of scientists who reveled in their well established—and closed—society. For example, *Science* reported, some said the book could not be true because "the scientists involved surely would want to publish" and because "no one working in any related field had heard anything about it." The more serious refutations included the obvious questions of whether the work in frogs could possibly be extended to clone a human being.

Newsweek reported in its March 20 issue, "Some noted researchers, including a group from Stanford University, were so apoplectic that they refused even to talk about Rorvik's claim for fear of giving the book free publicity." Clement Markert, an eminent molecular genetics professor at Yale University, told *Newsweek,* "I flatly disbelieve it." Beatrice Mintz, a leading mouse molecular biologist at the Fox Chase Cancer Center, said that Rorvik is "a fraud and a jackass."

Soon Rorvik critics were emerging from other quarters. Leon Jaroff, who had been Rorvik's editor when he worked at *Time* magazine, told *Newsweek,* "David is intelligent. David is a good writer. David is a little strange."

Scientists sought to soothe public fears. There would be no reason to clone humans, *Newsweek* said. After all, noted Peter Hoppe, a molecular geneticist at the Jackson Laboratories who would later emerge at the center of another cloning controversy, what would be the point? "We can't learn anything scientifically from human cloning," he said. Moreover, a person is more than genes alone. "If Hitler were cloned and distributed around the United States, there probably would be some very nice people," Markert said.

But some scientists decided that even if the book was not true, the time was nigh to prevent it from ever coming true. So without even seeing the book, three molecular biologists, Jonathan Beckwith of Harvard University, Ethan Signer of the Massachusetts Institute of Technology, and Liebe Cavalieri of Memorial Sloan-Kettering Cancer Center, joined with the People's Business Commission. This Washington lobbying group, run by persistent gadfly Jeremy Rifkin and his colleague Ted Howard, filed a Freedom of Information Act request and asked a federal court to expedite it, insisting on seeing what grants related to cloning had been awarded by the National Institutes of Health, the National Science Foundation, the Central Intelligence Agency, the Department of Agriculture, and the Department of Defense.

Beckwith said, "Even if this is a hoax, there is a good chance that human cloning is not too far off. Some public discussion should take place. We should ask whether there shouldn't be some legislative action."

And, he warned, "We'll wake up one day. Maybe it's not happened this time. But next time or the time after that, we'll find that we really have created a monster we didn't intend to create."

Congress planned to hold a hearing. A spokesman for the health subcommittee of the House of Representatives said, "If 99.9 percent of scientists are wrong and the story is true, there should be an open hearing to lay the issues out before the Congress and the public."

On the eve of his book's publication, Rorvik emerged to answer reporters' questions. He argued that no one could possibly make up so strange a story—an aging millionaire? A sixteen-year-old girl on a tropical island who falls in love with him and bears his clone? He said no serious science writer would risk his career by inventing such a plot and then passing it off as real. "Put yourself in this position. Would you dare risk writing such a story? In effect, you're jeopardizing your entire career." In response to demands from skeptics that he produce Max, Sparrow, and the clone, Rorvik said he was trying to persuade Max to come forward and to introduce himself and his clone to the world, but that Max refused. He added that Max welcomes the skeptics because they help assure him anonymity. "Every time someone cries hoax, he's delighted."

Rorvik had his defenders. Max Lerner, in a column in *The Washington Post,* said, that "at the risk of being a laughingstock for the rest of my life," he believed Rorvik's story. "Sometimes life imitates bad art," he concluded.

On May 31, 1978, Congress held its hearing on "the area of science most properly termed 'cell biology,' 'cloning' being one aspect of the general area." Rorvik did not show up. Paul Rogers, the congressman from Florida who chaired the Subcommittee on Health and the Environment of the Committee on Interstate and Foreign Commerce, explained what happened: "This committee has been in contact with him for more than two months. Originally, hearings were scheduled for April 21, and David Rorvik agreed to appear at that time. Due to what he indicated to this committee were personal health problems, the April 21 hearings were postponed and rescheduled for today. Once again, he agreed to appear.

"On the fifteenth of this month, he sent a telegram to the committee, informing us that he was extending his promotional tour for his book in Europe and that despite our previous agreement he would be unable to attend today's session. The committee wired Mr. Rorvik and urged that he attend as agreed, but we have heard nothing from him in response to our wire."

The scientists who testified, however, had plenty to say.

Beatrice Mintz said Rorvik's book "unquestionably is a work of fiction," adding, "It is nevertheless mildly amusing, although in ways not intended by the author. The amusement derives, for any qualified scientist reading the book, from the numerous scientific boners and errors with which Rorvik quite unwittingly discredits not only himself but also his so-called scientist, Darwin, who has reportedly accomplished this human cloning."

Robert Briggs, the man who, with Thomas King, first cloned a frog from immature frog cells in 1952, noted at the meeting that no one had ever cloned a frog that grew to adulthood, to say nothing of a mammal, from the highly specialized cells of an adult. He said that since cloning experiments, involving thousands of attempts in frogs alone, had failed, he suspected that "cloning in man or any other animal is not just a technical problem to be solved soon but may, in fact, never occur."

Clement Markert of Yale, who had said he flatly disbelieved Rorvik's tale, told Congress, "Our actual experience so far does not suggest that cloning with adult nuclei is possible, and, as has already been pointed out, many thousands of efforts have already been made in amphibians to test this possibility, and they have all failed."

Andre Hellegers, who directed Georgetown University's Kennedy Institute for the Study of Human Reproduction and Ethics, said he greatly enjoyed reading Rorvik's book. "I enjoyed it particularly because I knew it to be nonsense and I consequently had a grand time picking up the clues that would tell me it was nonsense. It was like reading Agatha Christie, but instead of proving who did it, you prove who did not do it." In short, he told the committee, "I entitle this book not *In His Image: The Cloning of a Man* but *In His Image: The Clowning of a Man,* meaning Mr. Rorvik."

James Watson, who just a few years earlier had written an article for the *Atlantic* saying it would soon be possible to clone a human and who had testified before Congress that cloning was imminent, now denied it in an interview in *People* magazine, published in 1978.

"When might we see the cloning of a man?" *People* asked Watson. His reply: "Certainly not in any of our lifetimes. I wouldn't be able to predict when we might see the cloning of a mouse, much less a man." Asked if he would ever clone a person, Watson vehemently replied that he could not imagine it. "What's to be gained? A carbon copy of yourself? Oh, if the Shah of Iran wanted to spend his oil millions on cloning himself, that's fine with me. But if either of my young sons wanted to become a scientist, I would suggest he stay away from cloning. There's no future in it."

Yet Rorvik's book became a bestseller—on the nonfiction list— in the United States and in England.

Mintz told me that she blames the media. "It was on the front page of the newspapers, it was in all the newscasts," she said. "It was just sufficiently improbable yet sufficiently alarming that it got a life of its own. How can you keep a thing like that from being a bestseller?"

And of course, some popular articles were careful to keep the door open to the possibility that Rorvik was telling the truth. An article in *Ebony,* for example, fanned the flames. " 'Bunk,' say numerous scientists. But some, recognizing the recent advances made in cloning techniques involving lower life-forms, do not so easily disclaim Rorvik's report," the story said.

But the media do not operate in a vacuum. In part, their interest and the interest of the public may have been sparked by the remarks of respected scientists like Jonathan Beckwith. He was, after all, a Harvard professor, and what he said made people curious enough about the book to buy it. When Beckwith hinted that there was some slight chance, incredible as it might seem, that Rorvik was telling the truth, it might have been hard to believe that the book was absolutely, indisputably, a hoax. Or perhaps it seemed that scientific skeptics like Beatrice Mintz were a bit too high-handed, that they protested too much.

The Rorvik phenomenon ended two years later. The way it ended, and the final reactions of scientists, can best be understood in the context of the increasing questioning of science in those tumultous years.

Rorvik's book appeared when the public's growing unease with science had swelled into an active distrust that persists today and that resonates in anguished debates over cloning. Many fear that scientists will secretly, or not so secretly, clone a human being, whether or not it is legal and whether or not the public approves.

The science debates that occurred when Rorvik's book was published marked a profound shift in public opinion. Granted, some questions about science had arisen in the 1960s. In 1962, there was the classic book by Rachel Carson, *Silent Spring,* that spawned the environmental movement. Carson warned that the wanton use of pesticides like DDT were denuding the earth of all its creatures. She exaggerated, to be sure, stating, at one point, that even the American robin was on the verge of extinction. But she changed the world, alerting the public to the fact that the chemicals concocted by industrial scientists might, in the end, be a Faustian bargain for our species.

In 1968, Paul Erlich's bestselling book *The Population Bomb* was published, a polemic that flatly asserted that the next decade would see worldwide famines. He was wrong, of course, but his book encouraged the public to believe that the best times might be behind us and that science might be powerless to pull us out of our looming troubles.

But the real science backlash did not begin until the 1970s. Before then, laboratory scientists, who had taken us to the moon and who were inventing miracle drugs and vaccines, still had a heroic image. It was not until the 1970s that a conglomeration of factors, including a movement by scientists themselves to hold off on their own research, merged to make the public skeptical of the promises of science and scientists and, perhaps, ready to believe a man like Rorvik who claimed that the scientists were wrong—that Max really had had himself cloned.

The ethics movement of the 1970s helped alert the public to the possible misuses of science. And so did a debate over the morality of in vitro fertilization and the terrors of its possible uses. This was a debate that Rorvik himself had helped generate. In his article for *The New York Times Magazine,* he wrote of the coming transformation of reproduction: Babies will be conceived outside the womb, women will hire others to bear their children, women will buy eugenically perfect embryos to bear and rear as their own children. In fact, every one of these predictions has come true, but at the time, they seemed truly horrifying. Embryos are not, of course, bought and sold, but the medical costs of obtaining one—paying for an egg donor, paying for a semen donor, paying more than ten thousand dollars to have the embryo implanted in a ready womb—makes them very much an economic venture.

Adding fuel to the flames was the bitterness that flowed from the Vietnam War. The war had spawned a generation that was distrustful of the government and government-funded projects. The destructive power of its new weapons, like napalm, cluster bombs, and Agent Orange, graphically characterized by the news media, had turned many students and professors away from industrial science.

Scientists themselves took part in questioning their own endeavors. They went on a research strike in March 1969, protesting the diversion of research money to studies of weapons of destruction. Three senior scientists at the Massachusetts Institute of Technology—biologist Boris Magasanik, chemist John Ross, and physicist Victor Weisskopf—wrote a letter to the journal *Science* explaining that they wanted "a public discussion of problems and dangers related to the present role of science and technology in the life of our Nation."

The first Earth Day was proclaimed in 1970. James Watson wrote that, at the time, "many academics thought that science was already out of control" and so they greeted Earth Day "as a watershed marking the moment when we stopped moving ahead willy-nilly without thinking through the consequences."

Part of the student movement of the early 1970s was a drive to change the nature of science, to turn it away from a focus on esoteric questions in molecular biology and toward a more populist enterprise that was concerned with public health and with the eradication of diseases, such as cholera and malaria, that strike in countries where people are desperately poor. Science students and young scientists flew to Cuba to teach or to help harvest sugar cane. Universities added courses on the social responsibility of scientists. Disaffected scientists formed a group, Science for the People, with its symbol of a laboratory flask and, behind it, a raised fist, that was devoted to exposing the underbelly of corporate research and much of the research at universities, which the group thought was heedless of the real needs of the world's population.

But the event that may have tipped the scales was a decision by leading scientists to impose a research moratorium upon themselves, for fear of unleashing catastrophe upon the world.

For the first time, biological scientists had become frightened by their own work and its implications. So when cloning emerged as an issue, through Rorvik's book and the debate it generated, it became part of that groundswell. In the 1960s, some intellectual leaders used cloning as a metaphor for the promise of science to allow humans to control their destiny and their evolution. By the end of the 1970s, in the context of the new movement to contain the awesome powers of biological scientists, cloning became a metaphor for the temptation of scientists to play God.

The scientists' movement began when researchers created tools that allowed them to isolate genes, strings of chemicals that are far smaller than anything that could be seen even with an electron microscope, and to move them from cell to cell. As the work progressed, each discovery was applauded, yet tinged with a shadow of fear.

In 1969, James Shapiro, then a rising star at Harvard University, announced that he and Jonathan Beckwith had isolated the first gene—a bacterial gene that microorganisms use to digest milk sugar. The feat was trumpeted on page 1 of *The New York Times* as a herald of a new genetic age. But, Shapiro said, the discovery was so awesome, and its ramifications so frightening, that he was retiring from science. He left for Havana, where he taught genetics. "It was one of those things you do with youthful abandon," he told me in 1992, when I interviewed him for *The New York Times*. After two years in Cuba, Shapiro returned to the United States. He is now a professor at the University of Chicago, still studying bacteria. He said, in retrospect, that he would not repeat the experience of fleeing to Cuba, but he refused to elaborate and declined to discuss that episode in his life.

The real terror arose, however, from discoveries that genes could not only be isolated but also could be moved. This meant that molecular biologists could create genetic combinations that had never been seen before in nature.

"Our first reaction was pure joy," Watson wrote. But soon, many scientists came to the sobering realization that these techniques were different in kind from anything that had come before. Now molecular biologists could add genes to bacteria that might make them impervious to any antibiotics humans could devise. It was conceivable that scientists, in all innocence, might create a pathogen that would wipe out life on earth.

As early as 1969, Joshua Lederberg, the Nobel prize–winning molecular biologist who had spoken out about science in his columns for *The Washington Post*, mapped out, in one of his columns, a rational argument to calm the worried public. Repeating the sorts of reasons he had used to argue that cloning would not be a horror, he told readers that instead of being frightened of molecular biology, they should view it as a way to substantially improve our species.

"Some of the more panicky reactions to genetic engineering and their characterization as 'tampering,' are, no doubt, very closely akin to the anxiety and derision that greeted Darwin's formulation of man's evolution from ape-like forerunners. How would an ape-prophet's relatives have greeted his predictions about the upsets their species would soon experience?" Lederberg wrote.

For many, the moment of insight came in the summer of 1971 when Janet Mertz, a young molecular biologist at Stanford University, announced that she planned to join DNA from a tumor virus with that of a bacterium that inhabits human intestines. She was speaking at a meeting at the Cold Spring Harbor Laboratory in Long Island, a private lab at the site of an old whaling village, which is devoted to cancer research. Every summer the lab holds meetings at which molecular biologists describe their latest results and what they plan to do.

Robert Pollack, a molecular biologist at Cold Spring Harbor, heard Mertz talk and was alarmed. The experiment involved so-called recombinant DNA, meaning DNA from different sources that was to be recombined. The tumor virus in question was originally isolated from monkeys, where it appears harmless. But the virus caused cancers in rodents, and scientists disagreed about whether it could cause cancer in people. Renato Dulbecco, a Nobel laureate tumor virologist, had said he would not hesitate to drink a solution of the virus. Others, like Pollack, feared that scientists like Dulbecco were carrying bravado too far, and that with no proof that the virus was harmless, we should consider it dangerous.

When Pollack heard Mertz speak, he envisioned the worst—a common intestinal bacterium, modified so that it caused cancer, let loose upon the world. So he called the head of Mertz's laboratory, Paul Berg, and pleaded that the experiment not be done. Berg was taken aback. "He was absolutely dumbfounded, as far as I could see," Pollack said. "I must have sounded like somebody coming and saying, 'God will punish you.' But when Berg asked some of his colleagues what they thought, they told him they agreed that the experiment ought not be done. And Berg changed his mind.

Wallace Rowe, a molecular biologist at the National Cancer Institute, remarked, "The Berg experiment scares the pants off a lot of people, including him."

From there, the debate accelerated. The question was, Should scientists impose rules upon themselves, voluntarily setting limits on their research? Some scientists argued that it was only right that they impose controls on themselves. Others decried the rush to regulate. The arguments lasted through most of the 1970s and were impassioned and divisive. The nation's leading scientists squared off in virtual phalanxes of *Who's Who.*

The debate thundered through the pages of *Science,* a publication that is read by hundreds of thousands of scientists. In 1976, Erwin Chargaff, a biochemist at Columbia University whose research on the chemistry of DNA in the late 1940s and early 1950s laid the foundation for James Watson and Francis Crick's discovery of DNA's structure, wrote: "Have we the right to counteract, irreversibly, the evolutionary wisdom of millions of years, in order to satisfy the ambition and curiosity of a few scientists?"

A few months later, Philip Siekevitz, a biologist at Rockefeller University, also wrote to *Science,* pleading for scientists to voluntarily restrain themselves. "Are we really that much farther along on the path to comprehensive knowledge that we can forget the overwhelming pride with which Dr. Frankenstein made his monster and the Rabbi of Prague made his Golem? Those who would answer 'Yes,' I would accuse of harboring that sin which the Greeks held to be one of the greatest, that of overweening pride. Like the physicists before us, we have entered the realm of the Faustian bargain, and it behooves all of us biologists to think very carefully about the conditions of these agreements before we plunge ahead into the darkness."

Physicist Freeman Dyson of the Institute for Advanced Study in Princeton, defended the biologists who wanted to use their new tools. He had come from the field that, as J. Robert Oppenheimer, the father of the atomic bomb has said, had "known sin." Their work had been used to create the bomb. But looking at biology from afar, Dyson saw that the greatest harm scientists could do to themselves might be to stop research that promised to benefit humanity.

"I claim that the exploitation of recombinant DNA may lead to an understanding, and conceivably to a cure, of cancer. It may lead to the creation of improved food plants which could save hundreds of people from imminent starvation. It may lead to the creation of energy crops which offer benign alternatives to nuclear fission and fossil fuels." Although there may be hypothetical dangers, he added, "let us not leave the starving millions of humanity out of the account when we balance the dangers against the benefits. It is perhaps not irresponsible, but rather an act of enlightened courage, to expose ourselves to an unknown risk of dangerous epidemics in order to give ourselves a chance of lifting some hundreds of millions of our fellow humans out of the degradation of poverty."

Some journalists tried to soothe the public. Horace Freeland Judson, a journalist who was writing a book on molecular biology at the time, wrote in *Harper's* magazine in 1975 that fear of science had gotten out of hand. He attributed part of the fear of molecular biology to "the feeling that this is the second time around, that the last time the ethical choices were muffled, and so the bomb was built." And yet, he asked, after musing over the possibilities of cloning and of moving genes from cell to cell, from organism to organism, "What are we afraid of?"

In June 1976, the National Institutes of Health issued elaborate research guidelines that would apply to all scientists who received its grants, insisting on strict measures to contain bacteria whose genes were being manipulated. But the University of Michigan and the city of Cambridge, Massachusetts, not content with the federal regulations, sought to impose their own. Soon New York State joined in, holding hearings on proposals to set guidelines that might be even stricter than the federal government's.

Once again, the feuding scientists emerged. Liebe Cavalieri of Sloan-Kettering Institution testified, "The risks of the research are worldwide and terrifying, the benefits are speculative." Nobel laureate James Watson, recanting his earlier statements that warned of the dangers of recombinant DNA, said, "I told Sargent Shriver that recombinant DNA is the most overblown thing since his brother created the fallout shelter debacle." Nobel laureate David Baltimore, who was a virology professor at the Massachusetts Institute of Technology, argued that scientists should police themselves. "I think that the scientific community, being as open as it is, and as self-critical, provides a better guarantee of safety than does any government regulation." But Baltimore's colleague at MIT, Jonathan King, argued that scientists are no more to be trusted to police themselves than the tobacco industry is to determine the dangers of cigarette smoking.

The debate continued into 1978. Lewis Thomas, the erudite doctor who wrote regularly for the *New England Journal of Medicine* on the wonders and beauty of science and medicine, wrote an article pleading for a continuation of recombinant DNA research. His article was published in *Science*. He wrote, "The pure research potential of the recombinant DNA technology is simply tremendous. It does not exaggerate the case to say that this may be the greatest scientific opportunity for biology in this century. . . . We need this new approach, not only for biology but for medicine itself.

"And yet, here we are, caught up in a public controversy in which the only issue being talked about seems to be the invention of monsters for their own sake, mini-Frankensteins, and it is even being made to seem as though this is really how the investigators engaged in work of this kind obtain their pleasure, like the mad scientists in their basement laboratories in grade B movies." He reassured his readers: "The workers in this field are not about to manufacture hybrid beings. They are trying to find out how things work."

James Watson, blunt as ever, commented in 1978 that bacteria created in recombinant DNA experiments were no Andromeda strain, the terrifyingly deadly bacteria posited by Michael Crichton in his eponymous bestselling book. "Ever since we achieved a breakthrough in the area of recombinant DNA in 1973, left-wing nuts and environmental kooks have been screaming that we will create some kind of Frankenstein bug or Andromeda strain that will destroy us all. Now we are threatened with a truly imbecilic law that could set back research for years," Watson said.

At the same time, science fiction writers had taken up the themes of science run amok: the dangers of recombinant DNA, ecological disasters, secretive scientists and the abuse of humans who served as research subjects. And, of course, they took on cloning. Book after book was published by well-known writers with vast popular appeal. Movies began to appear, taking up the same themes.

For example, Naomi Mitchison, the sister of British biologist

J.B.S. Haldane, took his optimistic notion that cloning could allow humans to control their own evolution and wrote a chilling book, *Solution Three*, that asked what would happen if this were attempted. In 1963, Haldane had declared that it would be possible to clone humans and that we would, of course, clone the best members of our society, enriching the population with geniuses and other exemplary beings.

In Mitchison's book, published in 1975, a devastating nuclear war has almost destroyed humanity. The survivors are racked by aggression and food shortages. So they decide to remake the human race by prohibiting sexual reproduction and cloning their best specimens: the man and woman deemed least aggressive. Naturally, disaster ensues. Those who are not clones rebel, violence erupts, even the clones turn out to be aggressive, and the solution to human evolution turns out not to be a solution after all.

In 1972, Gene Wolfe's novella, the first selection in his book *The Fifth Head of Cerberus*, explored another aspect of cloning. A father uses his son, a clone, to obsessively investigate himself. It is a story that combines the ethics scandals of the time, and the idea of the abuse of human subjects in medical experiments, with the fear that the clones would upend the natural order of personal relationships, making it impossible for the clone to truly be a son to his "father."

The most famous science fiction book about a clone was Ira Levin's 1976 bestseller *The Boys from Brazil*. Its premise was that the notorious Nazi doctor Josef Mengele tries to raise clones of Hitler, in the same sort of environment as the one that Hitler grew up in, and so to allow the Nazis to conquer the world. His theme included the chilling notion, repeated in Rorvik's book, that scientists are doing more than they tell the public. The cloning experiment, according to this book, was not so far removed from work that scientists were quietly doing all along.

The 1970s also ushered in the Star Wars movies in which audiences were informed that the wars between the Empire and the Rebels began with "the Clone Wars." *The Island of Dr. Moreau,* a chilling allegory of the godless hand of Darwinian evolution and the evils of science, was made into a movie in 1977, starring major stars like Burt Lancaster, Michael York, and Barbara Carrera. The evil doctor had tried to create chimeric animals that were humanlike, altering their brains and making them walk on two feet. In the end, the animals reverted to their true natures, and Dr. Moreau was killed by one of his own creations.

As the 1970s drew to a close, however, the recombinant DNA debates that had played such a central role in altering forever the public's image of biologists quietly faded away. The research guidelines imposed by the federal government slowly dissolved as it became clear that the hypothetical dangers of genetically engineered microorganisms, which had so frightened many scientists and the public, were overblown. By 1981, recombinant DNA had gone from being a threat to the continued existence of our species to a business opportunity. Genentech, a company founded by Stanford scientists who had helped discover the powers of genetic engineering, went public, valued by Wall Street at more than $200 million. The scientist-founders became millionaires.

As the genetic engineering threat slowly dissolved, there was no public reckoning. There was little debate over whether the extreme reactions of scientists and the public had been well founded, though overstated, or whether they were inappropriate and hysterical in light of what was known at the time. There was no lesson learned, no guidance for the next time a scientific advance caught us unawares. Instead, the transformation of recombinant DNA from the greatest threat since the atom bomb to a tool for the pharmaceutical industry occurred with little comment. "We had not put anything in place except as an experience," Paul Berg told me.

But the stain of the public dispute over the limits of science marked researchers. Biologists, like the physicists before them, had had to come out of their ivory tower to defend their actions to a public that was afraid of their glorious search for truth and beauty. Biologists had taken on the physicists' curse and become the group that was not entirely to be trusted. Although the public moved on to other concerns, many retained, in the backs of their minds, the image of the mad scientist who was ready to unleash such horrors as a bacterium that carried genes from a cancer virus without even thinking of possible consequences.

It was while the genetic engineering debate was raging white-hot, in 1978, when scientists were frightening themselves with the terrible portent of what they had done, and while the public was reading science fiction books and watching movies that raised fears of science and cloning, that Rorvik claimed that a scientist had secretly cloned a man.

He made that claim when the message—from fiction, from the recombinant DNA moratorium, from the Vietnam War backlash—was that scientists might not be so easily trusted and that science itself can easily run amok. The actual scientific reasons that made cloning seem so impossible were technical, sophisticated, arcane.

So it was perhaps not so surprising that David Rorvik's book found an audience. The only real surprise was how vehemently some scientists fought back, and how in their fighting back they took the strongest possible position—that cloning was not even on the horizon.

On July 11, 1978, just three months after Rorvik's book was published, J. Derek Bromhall, a British scientist, filed a $7 million libel suit against Rorvik and J. B. Lippincott, Rorvik's publisher. Bromhall, an embryologist at Oxford University, alleged in his complaint that the book defamed him by quoting from his studies with rabbit cells "so as to create the impression that Bromhall was cooperating or in some way had helped and was vouching for the accuracy or credibility of the book." Bromhall wanted a court order forcing Rorvik and Lippincott to admit that "the book is a fraud and a hoax, that it is fiction and that no cloned boy exists."

Bromhall said he had, in all innocence, sent Rorvik a nine-page summary of his doctoral thesis when Rorvik had written and indicated that he wanted to know more about Bromhall's work for an article or book he was writing. When Bromhall saw *In His Image,* he noticed that he and his work were cited. Rorvik referred to an "Oxford scientist" and identified him in a footnote as being Bromhall. He also described Bromhall's work in detail and included Bromhall's work in the bibliography of *In His Image.*

The passage that so inflamed Bromhall read: "Darwin, who by this time had enjoyed at least three glasses of wine, said that, in his opinion, no one would match his accomplishments for another ten years, at least. Then, embarrassed a bit by his own immodesty, he added that this would be so partly because others would be afraid to try.

"As a matter of fact, however, before the year was out, we would learn of the work of an Oxford scientist who had gone, if not 'straight for the throat,' then in at least an only slightly wavering line. This researcher reported in *Nature* that he had activated rabbit eggs with cold shock, used Sendai virus to fuse them with rabbit body cells, and had achieved, out of numerous efforts, four embryos that divided regularly at normal rates all the way to the morula stage, at which point they might conceivably have been successfully implanted, had the researcher been prepared to go that far."

Rorvik included a footnote that identified Bromhall and described, at length, his methods and his results.

Bromhall claimed that he had been defamed because Rorvik made it seem that his work was aimed at cloning humans and that he provided credibility to the story of Max and his clone.

In an opinion issued on October 1, 1979, a judge ruled that the book was a hoax but that Bromhall could not sue for libel. He wrote, "Accurate statements about the plaintiff do not become libelous merely because they are included in a book that is false in other respects."

Nonetheless, Bromhall persisted in suing for damages, arguing that his work had been wrongfully appropriated and that his privacy had been invaded when Rorvik used his name. The trial began on April 5, 1982, in the U.S. District Court in Philadelphia. Bromhall's lawyer, Arthur Raynes, showed the court a letter from Rorvik to Bromhall, questioning him about how to clone. The letter was dated five months after Max's clone was supposed to have been born. It was dated a year after Rorvik supposedly agreed to help Max clone himself.

Just three days later, but four years after *In His Image* was published, the lawsuit was settled. Lippincott said it would admit the book was untrue, would apologize to Bromhall, and would pay Bromhall a sum that was alleged to be about $100,000. It was the first time a publisher had been taken to court and the first time a court had declared a book to be "a hoax and a fraud."

Rorvik had earned $390,000 from his book, his publisher said, and Lippincott made $730,000. Lippincott would pay half of the settlement from its insurance and the rest from money it had withheld from Rorvik's royalties in order to cover legal expenses.

The apology to Bromhall, from Barton H. Lippincott, the chief executive officer of Lippincott, said in part: "Lippincott concedes it now believes the story to be untrue. Lippincott acknowledges that Dr. Bromhall did not consent to the inclusion of his name or his research technology in the book and also acknowledges that Dr. Bromhall was never engaged in, or attempted to engage in, or advocated the cloning of a human being. We regret any embarrassment, humiliation, or other injury."

Rorvik, however, never admitted that his story was untrue.

It had been "a bad episode for serious science," Mintz said. But scientists had, in effect, won this one. In the end, the public lost interest in Rorvik's claims. His book sold 95,000 copies in hardcover. Optimistic about its prospects, Pocket Books printed

675,000 paperback copies. But they failed to sell. "It was a big stiff," said Richard Clark, a spokesman for Pocket Books, in 1980. Rorvik faded into the background. The cloning scandal dwindled away.

Yet many scientists, in denying Rorvik's tale, had gone one step farther. They had adamantly insisted that cloning was not even on the horizon, that it could not be done in the foreseeable future, if ever. These scientists were soon to be rocked by a fantastic claim, by one of their own best and brightest, that a mouse had been cloned.

6

THREE CLONED MICE

The cloning of mammals, by simple nuclear transfer, is biologically impossible.
—JAMES McGRATH and DAVOR SOLTER
Science magazine,
December 1984

Sometimes in science a single failed experiment can set a field back for years, or even divert researchers' attention from a problem. It can tell earnest young people that their time is better spent elsewhere and it can signal to granting agencies that money poured into some projects is money wasted. So it was with cloning and with a strange roller-coaster ride that began with the news that three mice had been cloned and that ended, a few years later, with a report that not only had no mice been cloned but no mice could ever be cloned—and neither could any other mammal.

Within a few years from the end of the 1970s scientists went from believing that cloning would be feasible and interesting to thinking that it was a scientific dead end. The cliquish community of leading scientists who gathered at scientific meetings and, like the popular kids in high school, were objects of admiration and imitation, followed the claim of cloning with bated breath and its ultimate disparagement with mixed emotions—disappointment that the spectacular report might not be true tinged with a gossipy fascination with a tale of the rise and fall of a once mighty scientist.

In the end, the course of cloning research was changed. It moved out of the mainstream of science, away from the well-funded molecular biology labs and away from the renowned heroes of science whose pronouncements on what was important, or even interesting, carried the weight of received truth. Cloning became a pursuit of those who worked on the edges of science. It was relegated to those who worked with farm animals, and whose papers, even when they appeared in leading journals, often were not read by the scientific elite.

Few who were part of the past or future of cloning were untouched by the tale of the cloned mice. This was not a saga that unfolded in the pages of obscure journals or that involved scientists who came out of nowhere, made extravagant claims, and then were never heard from again. It involved a scientific superstar from a major lab whose papers were universally known by developmental biologists, by cell biologists, and by molecular biologists, and who had been on such extensive lecture tours of scientific meetings and labs that many leading scientists, and many who were later to become leaders, met him or heard his mesmerizing speeches.

The story began in the late 1970s when biologists were falling over themselves showering accolades on a German researcher, Karl Illmensee. He was trim and handsome, with a firm jaw and a military bearing, and the rumor was that he had golden hands. He could make an experiment work by the sheer virtuosity of his technique. Unlike other scientists, who would specialize, then superspecialize as they burrowed into a research question, Illmensee was versatile. First he worked on fruit flies, then mice.

And then, in the summer of 1979, he announced his most astonishing result of all to a gathering of the world's most distinguished
scientists. He had, he said, cloned three mice.

His announcement came at a time when the best scientists in
the world had despaired of such an achievement. One after another
had tried, and failed, to repeat with mice, the favorite laboratory
mammal of developmental biologists, what Robert Briggs and
Thomas J. King, did in frogs in 1952.

It was not just idle fancy that made researchers turn to mice.
If cloning was going to work in a mammal, mice seemed the
ideal animals to clone. Mice are mammals, like humans, and have
almost every gene that a person has—a mouse, in fact, is essentially a human with its genes rearranged. Moreover, mice breed
quickly and have huge litters. Finally, scientists have spent decades
creating the perfect strains of mice for research. Since the turn of
the century, scientists have been collecting and creating inbred
strains of mice that have been mated to one another so often that
the mice in each strain are genetically identical. If you wanted to
study a mouse whose blood pressure rises when you feed it large
amounts of salt, you could order the strain and get as many
identical twin mice as you desire that share a gene that makes
them salt-sensitive. So many molecular biologists make their careers studying the genetics of mice that the small rodents have
virtually taken over mammalian molecular genetics. You will almost never find a scientist studying molecular genetics in a squir-
rel, for example, or a vole. Larger animals were mostly interesting
for agriculture; molecular biologists were more interested in the
fundamental questions of life than in making better sheep, cows,
pigs, or goats. But every attempt to clone mice had failed.

It was excruciatingly difficult to try to exactly replicate the
frog experiments since they did not involve the fusing of whole
cells but instead involved the direct transfer of a nucleus from
cell to egg. Transferring the nuclei of microscopic mammalian cells
seemed technically impossible. But J. Derek Bromhall, the Oxford

scientist who had sued science writer David Rorvik, did manage a partial victory, using rabbits. He managed to suck the nuclei out of rabbit eggs and inject nuclei from early-stage rabbit embryos into the eggs, but the embryos failed to develop. They grew to about eighteen to twenty-four cells, then died, floating in shallow glass dishes in the laboratory, immersed in the thin broth of water, salts, amino acids, and sugars that was meant to nourish them until they were large enough to implant in the uterus of a female rabbit.

Perhaps Bromhall's method was inadequate, the instruments too blunt, traumatizing the tiny, delicate rabbit eggs.

But when scientists tried another method, it did not work either. First Bromhall, using rabbits, and then others, using mice, attempted to fuse whole cells with eggs, using an inactivated virus, called Sendai, that melds cells together. With the failure of the virus method, some scientists suggested that the virus itself fatally injured the cells.

Others said perhaps there was a real species barrier: An experiment that had worked two decades previously in frogs would never work in mammals.

So no one expected Illmensee's stunning achievement.

Illmensee's cloning was not the sort that Ian Wilmut announced when he cloned Dolly. It was much more limited in scope, involving cells from very early mouse embryos. When Wilmut cloned Dolly, he cloned her from cells from an adult, creating an identical twin of an animal that had already been born and had grown up, one whose features were known, whose nature was formed. Cloning cells from an embryo was not quite as evocative. After all, the animal would be an identical twin of what a microscopic ball of embryo cells would have been if they had been left alone and allowed to develop. If you were interested in duplicating an individual, who knew what sort of an individual an embryo would have grown up to be? But the cloning of a mouse embryo was astonishing nonetheless because no one had ever gotten that experiment to work, until Illmensee.

The experiment that Illmensee described was intricate, and seemed unbelievably difficult. He worked with Peter Hoppe, a promising young scientist at the Jackson Laboratory in Bar Harbor, Maine, a research institute incongruously nestled in the pine woods of the summer resort on Mt. Desert Island, a monotonous five-hour drive from the Maine border. The Jackson Laboratory is the world's only nonprofit research institution that was established solely to study the genetics of mammals. Over the years, it evolved into a center devoted exclusively to mice, keeping a mouse repository and sending out inbred strains of mice to researchers who requested them. Its director, Kenneth Pagen, told me that the lab now sends out half a million inbred mice annually.

The cloning experiment began when Illmensee and Hoppe put male mice into a cage with receptive females, allowing them to mate. Four days later, they flushed microscopic embryos from the uteruses of the females. At this stage, the embryo is still a tiny clump of about forty cells, looking nothing like a mouse. But the cells already have begun to arrange themselves, forming an inner cell mass—a ball of cells that will become a fetus—surrounded by a shell of cells that will become a placenta. Illmensee and Hoppe dissected the inner cell mass away from the rest of the embryo and then dissociated the ball of cells by adding an enzyme that dissolves the biological glue that holds cells together.

Their goal was to transfer the nucleus of one of the cells, which contained the genetic blueprint for the embryo cell, into an egg whose nucleus had been removed. So their next step was to shatter an embryo cell and fish out the cell's nucleus from the cytoplasmic muck that surrounds it. Illmensee said he sucked the embryo cell into a pipette that was too small to hold the entire cell. It was like trying to vacuum a beach ball. The cell was so delicate and the pressure on it was so great that, eventually, it would break apart. Its tiny nucleus, however, which was encased in its own jellylike membrane, would remain intact inside the slender pipette.

Illmensee then injected the nucleus of the mouse embryo cell into a newly fertilized mouse egg. As soon as he had gotten the nucleus in, he used the same pipette to draw out the egg's own genetic material. In this way, he said, he avoided poking the delicate egg more than once.

Finally, Illmensee and Hoppe let the cloned embryos grow in the laboratory for several days before implanting them into the uteruses of female mice, where they developed into normal mice that were clones of the embryos that Illmensee started with.

The experiment was a tour de force. It was the first time anyone had cloned a mammal. And even though the mammal was actually an early mouse embryo, not an adult mouse or even a fetus, a barrier had been broken.

"There was no reason to be skeptical," recalled Shirley Tilghman, a molecular biology professor at Princeton University. Illmensee, she said, "had a reputation for being a wunderkind." He had produced a "spectacular Ph.D.," working with fruit flies in the late 1960s. In those experiments, he had asked whether the cells in an egg "knew"—had biochemical instructions that specified— where they were supposed to be during development. It was a century-old question that had defied the greatest minds in science.

Illmensee had completed an experiment that seemed impossible. He had actually transferred cytoplasm, the gel that makes up the inside of a cell, from one cell of a fruit fly's egg to another. He could then ask, Did the cell, in its same position but with the cytoplasm from a cell that came from elsewhere, grow as if it were in its current position or as if it were in the position of the cell whose cytoplasm it now had? The answer, he discovered, was that the cell behaved as if it were from another part of the egg. He concluded that the cytoplasm contained instructions telling the cell where it was.

The fly experiment also was one that senior scientists had attempted, but failed to accomplish. "When Illmensee pulled it off, he got a reputation for having golden hands," Tilghman said.

From there, Illmensee went to the laboratory of Beatrice Mintz, the award-winning scientist at the Fox Chase Cancer Center in Philadelphia who studies mice.

In his work with Mintz, Illmensee reported that he had mixed normal mouse embryo cells with cancerous cells from a type of tumor that arises from embryos. The mixture of normal and cancerous cells, which formed what scientists call mosaic embryos, developed normally, and the tumor cells even became part of the testes and ovaries of the resulting mice. When Illmensee bred the mosaic mice, producing a second generation, he found that the genes from the cancer cells had been passed on. But the mice did not develop cancer. The cancer cells had reverted to normal. The results, Tilghman said, "were a big surprise" because, once again, this was an experiment that had eluded other scientists.

In 1977, Illmensee electrified molecular biologists when he announced that he and Peter Hoppe had created mice with a father but no mother and mice with a mother but no father. This experiment cemented Illmensee's already growing reputation and set the stage for his claim that he had cloned mice.

When an egg is newly fertilized, the genes from the egg are separated from the genes from the sperm. These two sets of genes, called the pronuclei, soon come together to form the complete genetic material needed to direct the embryo's development. But Illmensee and Hoppe decided they could exploit the fact that, for a brief time, the two sets of genes are distinct. They reported that they removed one of the pronuclei from newly fertilized eggs and then grew the eggs in a broth containing an enzyme that encouraged the pronucleus of an egg to duplicate itself. Eggs with a single set of egg genes would become eggs with a double set of those genes—in essence, embryos with a mother but no father.

Eggs with a single set of sperm genes would become embryos with a father but no mother.

Illmensee and Hoppe said that they had created five fatherless mice and two motherless mice. These experiments were dazzling, scientists thought, seemingly a violation of the laws of nature. Others, including leading scientists like Clement Markert of Yale, had tried to make mice with two mothers or with two fathers, and failed.

Tilghman will never forget the reaction of the most renowned molecular biologists to Illmensee's report that he had made mice that had just one parent instead of two. She heard him talk about these experiments in the summer of 1977 at a Gordon Conference, one of the most exclusive scientific meetings. Every year, the scientific elite gather at these conferences, held at New England prep schools, where they live in spartan dormitory rooms, sleeping on narrow swayback beds, and spend their days in classrooms, where they hear reports of the latest, as yet unpublished, discoveries. To prevent grandstanding, to foster an atmosphere of collegiality, and to allow scientists to speak freely, no members of the press are allowed to attend. Since the schools have limited space in their dormitories, enrollment is limited to no more than about 120 scientists, and the scientists must apply and be accepted in order to attend.

Tilghman was a postdoctoral student, still learning the ways of science, when she attended that conference. At lunch in the Princeton faculty club, on a bright day in early spring of 1997, twenty years after the event, she told me what happened, her eyes widening at the very thought. It was, she said, "the only time in my career that I ever saw an audience applaud in the middle of a scientific speech—the only time I have ever seen that in my life."

So when Illmensee, who had become a professor at the University of Geneva, said he and Peter Hoppe had cloned three mice, many of the most skeptical of scientists were prepared to believe that if anyone could pull off this experiment, it was Illmensee.

*   *   *

Rumors of the cloning started to create a buzz among scientists, but the public was unaware of what had happened until Sunday, January 4, 1981, when *The New York Times* printed a page 1 story by Walter Sullivan announcing the feat. FIRST CLONING OF MAMMALS PRODUCES 3 MICE, the headline said. At the start of the story, Sullivan explained its significance. "Earlier reports that mammals and even a human being had been cloned have never been authenticated or taken seriously. This is the first such report to be accepted by the referees of a leading scientific journal."

The next day, newspapers around the world blared the astonishing news. Magazine writers also took up the gauntlet. *U.S. News & World Report* announced, "Scientists now are within one step of what once was termed science-fiction fantasy: the cloning of a mammal. *Newsweek* reporter Sharon Begley asked, "Now that a mammal has been cloned, can a man be far behind?"

Charles Krauthammer, writing in *The New Republic,* tried to soothe the public's fears. The *New York Times* story, he wrote, "sent many scurrying for genetic bomb shelters." But, he cautioned, the mice were cloned from early embryos, not adults. "On the whole, the three mice represented a considerable technical achievement, but, for mad scientists, rather a boring one." The point of cloning, he said, "is to clone Mozart. It's no fun cloning an embryo if you don't know it's going to be Mozart."

Illmensee and Hoppe's paper appeared in the January 1981 issue of *Cell,* a leading journal for molecular and cell biologists, one that accepted papers only after its referees, respected scientists in their own right, had read the papers and satisfied themselves that the papers were close to flawless and of major importance. The three mice graced the cover of that issue of *Cell,* facing forward in a phalanx. Two of the cloned mice were gray and one was dun colored. Illmensee said he had cloned cells from embryos of gray and dun-colored mice. The eggs had come from black mice. And the surrogate mothers, who carried the clones to term, were white.

Illmensee soon was avidly sought out by other scientists, becoming a key player on the scientific lecture circuit. Like the golf or the tennis circuit in sports, scientists have their own circuits: an endless round of meetings and seminars in which competitive scientists receive publicity and establish their pecking order. Part of being well known in science is to be on the circuit. And much of being competitive in science involves being on the circuit. Virtually all scientists confess that they know who is on the circuit and who is off, who is clambering on and who has been dropped, and the best ways to get on in the first place. There also are minor circuits for scientists whose work does not qualify them to compete in the major leagues. "It's like celestial spheres with epicycles," said Ponzy Lu, a biochemist at the University of Pennsylvania.

Illmensee was on the major circuit.

Brigid Hogan, a developmental biologist at Vanderbilt University, said she was just moving into the field of mammalian development when Illmensee's results became known. "I was in London, at the Imperial Cancer Research Fund, when that work was coming out. Everyone was rushing around and inviting Illmensee to give talks. He certainly gave a very big lecture at the Imperial Cancer Research Fund, one of many" that he gave that year as the world's leading scientists deluged him with invitations to visit.

Keith Campbell, who was Wilmut's collaborator in cloning Dolly, heard Illmensee speak while he was a doctoral student in London, and will never forget the electric excitment his talk inspired. Steen Willadsen, a daring Danish scientist who was the first to clone a sheep from a sheep embryo cell, and whose work paved the way for Wilmut and Campbell, also heard one of Illmensee's lectures and was inspired enough to try Illmensee's methods.

And starry-eyed students sought him out.

Patricia Kahn, who is now a correspondent for the journal *Science,* said she first was drawn to Illmensee when she heard him lecture at Albert Einstein College of Medicine, where she was working on her doctoral degree in genetics. It was 1978, when Illmensee had just finished his tenure with Mintz. "He gave a seminar that was just mind-boggling," she said. "I decided on the spot that I wanted to work in this man's lab."

Illmensee was mesmerizing and attractive. He looked, Kahn said, like the Australian actor Paul Hogan, who played the starring role in the movie *Crocodile Dundee.*

Kahn visited Illmensee the next autumn, at the University of Geneva. "His lab was very new at the time; it was still small and it had a very nice atmosphere. I just loved it and he was delightful."

So, with a fresh doctoral degree, Kahn came to the lab in the fall of 1981, the same year that Illmensee's paper had appeared. By then, she said, Illmensee's lab had grown and Illmensee "was rolling. He was a very famous person."

But working in that lab was not the delightful experience that Kahn expected. Illmensee, she said, did not encourage the other scientists to work on a problem, as most lab directors do. Instead, each person was supposed to focus on a technique. Her impression was that the scientists were kept unaware of the larger picture. She felt they were like workers on an assembly line. Kahn's job was to grow cells from a type of human cancer called teratrocarcinomas. While working on her doctoral degree, she had perfected the skill of growing human cancer cells, but she had never intended that to be her life's work. The teratrocarcinoma cells were difficult to grow, however, and she worked on the project by trial-and-error, trying to coax the cells along. Yet Kahn did not even know—and Illmensee never told her—why he wanted her to grow these cells. She was twenty-seven years old, eager to do an interesting, challenging piece of research, and she was frustrated, irritated by her project. She had gone to Geneva to study development and had dreamed of learning Illmensee's method of nuclear transplantation, but she had become mired in this mysterious project, growing cancer cells.

But within a few months after Kahn arrived in the lab, one of the other scientists suggested a project that she was anxious to work on. It involved taking normal mouse embryo cells and trying to grow them in the laboratory. After she had gotten them established, growing well in dishes in the lab, the plan was for her to try to transfer their nuclei to other embryo cells. It was the perfect experiment, Kahn told me, because whatever the outcome, it would tell her something about how cells developed. And it was an opening into an exciting new project in the molecular biology of early development—the field she had come to Geneva to study.

She began attempting to grow the cells. It took a few months, but she finally saw the first glimmer of success, whereupon she rushed in to see Illmensee and to see if she could learn to transfer nuclei, a procedure that, she realized, would take months to master.

At that point, however, something peculiar happened. "It started to get strange," Kahn said. Instead of encouraging her enthusiasm, Illmensee "said there must have been a misunderstanding, that he personally would do the nuclear transplants and that he wasn't prepared to teach me." Moreover, Illmensee suddenly discouraged her from continuing the experiments altogether, for no clear reason.

"That was the beginning of the end for me," Kahn said. She started questioning the others in the lab and discovered, to her great shock, that no one else in the group had ever done a nuclear transplant and that no one had ever seen Illmensee do one. "Within a week, I left the lab," she said. "I felt there was something wrong there. I was having deeper and deeper misgivings."

Although Illmensee told Kahn that he did not care if she left his lab, he also told her "that he never wanted to hear that I uttered a bad word about his lab or he'd see to it that I never worked in science," Kahn said. She returned to the United States for a visit, but her former professors, still dazzled by Illmensee, did not want to hear her story. "Their reaction was distinctly uncomfortable. They said, 'Oh, that can't be. These are misunderstandings. Don't say those things or you'll bring trouble on your head,'" Kahn told me. She felt abandoned and betrayed.

Just before Kahn left Illmensee's lab, two young Americans studying in England came to visit Illmensee, also hoping to join his lab and learn nuclear transfers.

Elizabeth Lacy, a molecular biologist at Memorial Sloan-Kettering Cancer Center in New York, was enthralled when she first heard Illmensee speak. A postdoctoral student at Oxford, she said she was stunned by Illmensee's talk. Illmensee, she said, was a dynamic speaker, utterly convincing. Her husband, Frank Constantini, now a developmental biologist at Columbia University in New York, was trying to repeat Illmensee's work, with no success. He blamed himself. "I thought it was just me," he said.

Lacy and Constantini, in fact, decided they would move to Switzerland to work with Illmensee at the University of Geneva. They visited his lab over Christmas, when Illmensee had returned home, and had dinner with him at a small hilltop restaurant just over the border in France. Constantini even remembers what they ate. "For the first time, I had fennel," he said. "It was a revelation. I've cooked it ever since." Illmensee was charming. But something seemed amiss.

Lacy said she was uncomfortable when Illmensee said he would not teach them his technique unless they actually joined his lab. Other scientists routinely spent time demonstrating their methods to all who asked. And the dozen or so junior scientists in the lab seemed uneasy about Illmensee's work. "We didn't get the best vibes from the people in his lab," Constantini said.

Some senior scientists, too, were beginning to wonder whether something was not quite right. Mintz, who had had a bitter breakup with Illmensee, said she never saw any evidence of his so-called golden hands. "I realized almost immediately that he was a person with problems," she told me. "The first papers he published after he left here were just so patently impossible that I couldn't understand whether he had just managed to bewitch people in the field or whether they weren't taking those papers seriously." For example, she said, one paper involved the insertion of a human gene into a mouse. Illmensee published photographs of the mice with the added gene. "I was well familiar with mouse coat genetics," Mintz said. And one thing was perfectly clear to her. "The animals he had pictures of could not have resulted from the experiments he did."

Meanwhile, others were frustrated when Illmensee politely avoided demonstrating his techniques. Davor Solter, a developmental biologist working at the Wistar Institute in Philadelphia at the time, also found that he could not repeat Illmensee's work. He had wanted to clone ever since coming to the United States from Zagrev, Croatia, in 1973. But, he said, "it was hard to find anyone who was interested," and the "few small attempts" made by scientists "didn't lead to anything." Solter saw cloning as a way to get at the ultimate scientific question of defining the molecular basis of development. Does the genetic material of a cell change when a cell is specialized and, if so, can it be brought back to the state it was in when an egg was first fertilized? It was exactly the question that had tantalized the great German embryologist Hans Spemann decades earlier.

Finally, Solter told me, he found a postdoctoral student, James McGrath, who wanted to clone. "I was thinking about how one could do it," Solter said, when Illmensee announced his result. So Solter and McGrath abandoned their own idea of how to clone and adopted Illmensee's. But, Solter said, although they tried repeatedly for more than a year, they could not get the method to work.

When Illmensee visited the Wistar Institute, where Solter and McGrath were laboring away, Solter saw his chance to be taught by the master. "We said, 'Karl, we have been struggling for a year. Show us how to do it,' " Solter told me. And Illmensee did not refuse. But somehow, it never came to pass.

Over and over again, Solter's plans to lure Illmensee to the lab fell through. "We would collect the eggs, but he was too busy," Solter said. Every time Solter coaxed Illmensee into his lab, there was some reason why Illmensee could not demonstrate his method. He would say he was too busy just then, tomorrow would be better, or the pipettes were dirty or the equipment was not set up right. Finally, it was Illmensee's last day at the Wistar Institute. It was a Saturday, Solter recalled, and he nudged Illmensee into his lab, saying, "Now it's time to really show us."

The embryos and the eggs were ready for the cloning to begin. Illmensee sat down and started working. But, Solter said, "after two or three minutes, he said, 'The needles are very sticky,' and got up and left."

Solter and McGrath looked at each other. "I told Jim that he probably doesn't have the slightest idea of how to do it and probably never did," Solter told me.

At the same time, scientists also were trying, and failing, to repeat Illmensee and Hoppe's creation of mice with a mother but no father or with a father but no mother. "Those who contacted Illmensee all got this story about how you have to be extremely skillful, extremely good," Solter said. "I was a little bit suspicious. This is all nonsense. There is never anything in mammalian genetics that some can do and others cannot."

In the meantime, scientists in Illmensee's lab were growing leery. It began with subtle indications that something was amiss but it took years before mild wariness turned to outright disbelief.

Denis Duboule was a doctoral student in Illmensee's lab, arriving in 1977 just after Illmensee had accepted his position at the University of Geneva. "At that time he had a tremendous reputation," Duboule told me. But Duboule began to have doubts after a few years when he and another scientist in the lab, Kurt Bürki, noticed that "at a time when things didn't work that well in the lab, he was getting good results." He and Bürki began to look more closely at Illmensee's work, "in the first place, not with the idea that he was doing something wrong, but with the idea that we would like to understand what was going wrong for us," Duboule told me.

By that time, Bürki had been in Illmensee's lab for about three years. He had left his position at the Swiss Institute of Cancer Research because Illmensee, Bürki said, "had such a huge reputation." Bürki was a senior scientist, on Illmensee's level, who became personal friends with Illmensee and his family. "Our families met very often outside of work," Bürki said. In the beginning, he told me, he was completely won over by Illmensee—and so were the others in the lab. "We really thought he was a wonderful scientist."

Bürki became concerned about Illmensee's work when he proposed an experiment to Illmensee. It involved creating mice with just one parent. Illmensee "always had excuses not to do it," Bürki said. "So I started to do it on my own in his lab."

Illmensee had warned Bürki that the technique was very difficult. But when Bürki tried it, he soon discovered that it was actually very easy. He had no problem removing the nuclei from the eggs, as Illmensee had done. He had no problem keeping the eggs, which now had just one parent, alive in the laboratory. He had no problem transferring the embryos to female mice. But, he said, not a single live mouse was born. He had to wonder: If the technique went so smoothly, why did he not get Illmensee's result—a live mouse with one parent?

That same year, the lab had a problem with its water purification system and the finicky mouse eggs would not grow in their thin soup of water and nutrients. Except for Illmensee's eggs. "He claimed there was no problem," Bürki said. "He said he was successfully culturing eggs."

Finally, a student in the lab, Joachim Huarte, went to the incubator and looked in the test tubes that were supposed to contain Illmensee's mouse eggs. The tubes, he reported, were empty.

On Friday, July 11, 1982, Duboule said he noticed that the micromanipulator, the instrument used to transfer nuclei between mouse cells, had a cracked pipette attached to the microscope. Anyone who wanted to use the device would have to replace that pipette with a good one.

Illmensee said he had come in that weekend and used the micromanipulator in his experiments. But, Duboule noticed, the same cracked pipette was still there on Monday, and it was still in focus in the microscope.

In the meantime, Bürki said, Illmensee continued to work at night or on weekends, when no one was around. Illmensee never taught his methods to the others in the lab, and never demonstrated. "We saw him sitting at a microscope, but we couldn't check what he was doing," Bürki told me.

Bürki persisted. "When I was certain that he had cheated on the experiments, of course I have no big choice. I immediately realized that if he was cheating on these experiments it was highly likely that he had cheated before. Later on, people start asking you, 'Did you see anything?' Then I would have to say that either I was a fool and I didn't see anything or I would have to lie. I chose to make things clear. But we wanted to make things clear with him, not with the public, only with him."

"All these things together finally persuaded us that we had to talk to him," Bürki told me. "We didn't go public or to the university. We tried to talk to Illmensee, to persuade him that he should explain what he was doing. We had a feeling that something was wrong and he had to justify what he was doing."

But Illmensee was hostile, angry, Bürki told me. "When I tried to speak to him, he became quite aggressive and tried to isolate me and tried to persuade the others that he was right."

Finally, Bürki decided that he had to make some sort of statement. He chose his moment. On January 14, 1983, Illmensee was presenting new, unpublished results to a group of scientists at his university. When he finished, Bürki stood up. He stated that he and some of the others in the lab would not accept Illmensee's results.

The scientists in the audience were stunned. It was a statement that implied that Karl Illmensee, the leading light of the biology department at the University of Geneva, had falsified his results.

Several biology professors, including the head of the department at the university, began meeting with Bürki. A month later, Bürki gave a statement to the dean of the university, Hubert Greppin, describing his concerns. It was short, five and a half single-spaced pages, but direct.

Bürki wrote that while he was away on vacation in June 1982, Duboule and Huarte "had noticed discrepancies between statements made by Professor Illmensee and their own observations concerning the ongoing experiments." For example, he said, "on several mornings" Illmensee "presented females that had not been mated the previous night," yet claimed he had gotten embryos from the mice—an impossibility if the animals were not pregnant. Bürki wrote that there were test tubes in the incubators labeled with the same dates as the dates the mice had reportedly been mated. Those test tubes were supposed to contain fertilized eggs from the mice, but they were empty. In the freezer were "a large number" of micropipettes that were supposed to contain frozen embryos grown from those fertilized eggs, an impossibility, Bürki noted, if the test tubes in the incubator never contained eggs in the first place.

"Daily, we continued our observations without mentioning

it to Prof. Illmensee," Bürki's report continued. On the morning of July 11, Illmensee claimed to have transferred nuclei
from mouse tumor cells to fertilized eggs whose nuclei had
been removed. "As we had not observed any normally developing embryos in any of the recent culture tubes, Mr. Denis
Duboule and I told Prof. Illmensee about the discrepancies between his statement and our observations concerning the in
vitro culture of the eggs," Bürki wrote. Moreover, he added,
he and Duboule told Illmensee that it should be impossible to
grow mouse embryos in the laboratory because of the water
filtration problem. Illmensee assured him and Duboule that he
would prove to them that he could grow mouse eggs even
though Bürki and Duboule had been unable to.

So Illmensee began a new series of experiments in which he
transferred tumor cell nuclei into mouse eggs. Each day, he
showed Bürki and Duboule mouse embryos that were at the correct stage of development. He told them that these embryos now
contained nuclei from tumor cells and, as Illmensee had asserted
earlier, they were growing normally. And yet, Bürki wrote, he
and another scientist in the lab, Ulrich Petzoldt, also were trying
to grow mouse embryos but had been unable to keep them alive
beyond the four-cell stage. Moreover, he and Duboule noticed that
Illmensee had a collection of mouse embryos, at the appropriate
stages of development, hidden in another incubator. "Therefore,
in later discussions with Prof. Illmensee, we did not accept this
series as proof that Prof. Illmensee was able to culture eggs under
conditions where other experienced scientists were not able to do
so," Bürki wrote.

Bürki continued, describing "discrepancies" in Illmensee's laboratory notes and even noting that Illmensee appeared to have notes
describing experiments before they were done. These were a series
of experiments that involved cells that Bürki had grown. Illmensee's notes said he had done the experiments in April and

May 1982 when he had told Bürki to start growing the cells in July, after Bürki's vacation.

On May 17, 1983, Marco Crippa, the head of the biology department at the University of Geneva, and two other biology professors met with Illmensee to discuss the accusations that he had cheated. They did not, however, show him Bürki's statement. By the end of the meeting, Illmensee signed a statement, and the other professors countersigned it, saying, "Protocols of Dr. Karl Illmensee have been manipulated in a way which is contrary to the scientific ethics in some period of 1982." In a letter accompanying the statement, the faculty members wrote, "Dr. Illmensee clearly recognized having falsified ('faked') protocols including experiments that in reality had not been carried out."

The next day, Illmensee wrote a letter to the National Institutes of Health in the United States withdrawing a sentence from a grant application he had submitted in May 1982. It was a sentence claiming he had created mice by transferring the nuclei of tumor cells into mouse eggs.

At this point, the university was in an uproar, and news of the incipient scandal had spread from lab to lab in Europe, in England, and in the United States. But although newspapers like *The New York Times* and weekly magazines like *Time* and *Newsweek* had heralded Illmensee's claims to have cloned mice, they were strangely silent about the new charges that Illmensee's work might not be credible. A few short articles appeared but they hardly conveyed the dimensions of the controversy that was rocking the scientific community—and what it implied about the possibilities for cloning.

In August 1983, Illmensee consented to an exclusive interview with a popular German magazine, *Bild der Wissenschaft*. He categorically rejected the charges against him and agreed that he was especially gifted in the lab. "For example, when the concert pianist

Ashkenazy plays Prokofiev and you compare him with another pianist playing the same piece, one will hear big differences," he said. "I simply have qualities and abilities which I've picked up over the years that not every scientist has."

Asked if he himself did "the definitive parts of that experiment," Illmensee replied, "I did the important micromanipulations on the mouse egg myself. This is comparable to the surgeon, the definitive incision he does himself. My colleague, Peter Hoppe, was responsible for the cultivation and growth of the embryos."

Among scientists, the accusations were taking on the spectacle of a true scandal. Because the experiments that Illmensee had reported were so important and because the charges against him were so grave, the University of Geneva decided that it had to appoint an international commission to examine the evidence against him and to decide whether the charges had merit.

The five-member commission met first on August 22, 1983, taking depositions from the principals in the case. Illmensee brought in a lawyer and defended himself vigorously. Every time the scientists in his lab claimed they had caught him cheating, he provided an alternative explanation.

The mice that had not been mated yet that, presumably, had been the source of embryos? Illmensee said that they had been mated and that Huarte was just not experienced in recognizing that. After mating, female mice display a vaginal plug, a hard cream-colored pellet that looks like a piece of chalk and is formed from the seminal fluid of the male. A vaginal plug, explained Lee Silver, a mouse molecular geneticst at Princeton University, is completely obvious to anyone who looks. "You can't miss it," he told me. But Huarte, Illmensee said, had simply not noticed the plugs. Since Huarte had not kept a record of how many female mice he had checked, or on which days, Illmensee argued that Huarte's charges were not well grounded.

The cracked micropipette attached to the microscope that Ill-mensee said he used on the weekend when no one else was in the lab? Illmensee said he always left a pipette in the apparatus and so, after he had finished his experiments, he simply replaced the pipette he had used with the one that was there when he arrived. No one had recorded how many pipettes were on hand before and after that weekend, so there was no way to decide whether Ill-mensee was dissembling or that Duboule was mistaken.

The embryos that Illmensee claimed to have grown at a time when the water filtration problem seemed to have precluded others from growing embryos? Illmensee said he made no such claims, and there were no independent witnesses who could determine who was telling the truth.

The most damning evidence of all might appear to be Ill-mensee's statement, which he signed just after Bürki accused him of cheating. Throughout the day and evening of its first meeting, on August 22, the commission discussed what the statement meant. The professors who were with Illmensee when he signed it said it clearly was a confession, an acknowledgment that he had faked his results, and that Illmensee had agreed that that was what was meant when he signed it. Illmensee, his lawyer at his side, said that was not what he had meant at all. He had only meant to convey that he'd made mistakes in writing up his lab work, not in actually doing the experiments.

After a total of eight meetings, some lasting several days, this science court, in which renowned researchers served as judge and jury, issued its verdict. Illmensee, the commission found, seemed to have been extraordinarily sloppy in recording his experiments and in dating his lab notes.

The committee wrote that Illmensee had made an "unacceptably large" number of errors in his experiments, as evidenced by "corrections, confusions, and discrepancies" in his lab notes, particularly in June and July. "His method of recording," the group added, "is inherently prone to error and hard to check." The committee also criticized Illmensee's secretive ways and his failure to teach his techniques to his colleagues. Such behaviors, "while perhaps understandable in a man who prides himself on his personal contribution to pushing back the frontiers of knowledge, provide an obvious basis for arousal of suspicion among those with whom he works. When added to Illmensee's apparent inability to communicate with his staff, it proved to be a recipe for disaster." And so, the group wrote, "it is perhaps not surprising that some of Professor Illmensee's junior colleagues were led to suspect that some or all of his experiments were fabricated."

The committee did say that because of the sloppiness of Illmensee's documentation, the entire series of experiments under question is "scientifically worthless" and should be repeated "with full scientific rigor."

So, with the fraud charges unproven, Illmensee returned to his position as full professor in good standing at the University of Geneva. But he lost his research grants from the Swiss National Fund and from the National Institutes of Health. The National Institutes of Health also said that it would not give Illmensee any more money until he could validate his disputed results.

In the United States, many scientists felt the committee had cleared Illmensee. In Europe, Duboule told me, the opinion was different.

"There are several ways to read the commission's report and it is very clear that there are big differences between the American way and the European way to write these things," Duboule said. "People here are much less straightforward. All my American colleagues felt it had been totally whitewashed. Here in Europe, people tended to think it was a strong report."

Bürki and Illmensee had had to suffer the hell of each other's company throughout the year of Illmensee's investigation. "It was not a very nice atmosphere," Bürki told me, but they endured.

And when the committee decided that the fraud charges against

Illmensee were unproven, Bürki felt resigned. "We had to accept the committee's decision," he said. "We did not behave as policemen there. We did not set traps." Finally, Bürki left, taking a job at a Swiss drug company, Sandoz Pharmaceuticals, which is now called Novarsk, and there he remains, still working with mice.

Duboule did not fare so well. "I essentially was fired by the university three months before my Ph.D. defense," he told me. "I had no right to return to the lab. The university threatened us with, how do you call it? slandering? defamation?" Illmensee, he added, "was a very influential and famous person. They had a choice between a very influential professor and a student. I don't blame them."

It was a potentially devastating outcome for Duboule. The defense of the thesis is the last stage of the long, grueling process of getting a doctoral degree. Six years of work in Illmensee's lab, endless hours laboring over his experiments, and he was thrown out with no degree and barred from entering the lab.

But Duboule persisted. He managed to find another scientist who would sponsor him for his doctoral degree. And then, he said, he got a job in the laboratory of Pierre Chambon, a world-renowned molecular biologist at the university of Strasbourg and a member of the commission that investigated Illmensee. "That I could survive at all was exclusively due to Pierre Chambon," Duboule said. "Otherwise I would have been out." Now he is back as a full professor at the University of Geneva, where Illmensee once was. "So I think there's some kind of justice," Duboule said.

Peter Hoppe, in the meantime, also was investigated at Bar Harbor Laboratories, where he worked. Illmensee's troubles had spilled over onto him, casting doubt on the experiments that he and Illmensee had done together. The Bar Harbor investigation also could find no evidence of fraud, so Hoppe remained there.

But his life was ruined. "My reputation was tarnished," he said, and when he applied for grants to support his research, he was turned down. "The worst part for me was the grants. They gave me a score that said, 'Just get out of science,'" Hoppe told me. The Jackson Labs gave him a job creating genetically engineered mice for other scientists to use, but he no longer did research. Finally, in 1995, he retired, at the age of fifty-three. "I walked out with my head high," he said. But he is bitter. Illmensee, he said, "ruined so many people's lives, not only mine."

Illmensee remained at the University of Geneva until July 1985, when his professorship was up for renewal. In Swiss universities, professors do not have lifetime jobs, as they do in the United States, but instead must have their positions renewed every seven years. Before the university could decide whether to renew Illmensee's position, he resigned.

He ended up at the University of Salzberg in Austria, and he now works at a gynecology clinic there, doing in vitro fertilization and prenatal diagnoses.

In the meantime, Davor Solter and his postdoctoral student, Jim McGrath, had persisted in their attempts to transfer nuclei among mouse cells. After their experience with Illmensee's method, and after Illmensee evaded their assiduous attempts to get him to show them how he did it, they went back to the method that Solter had originally planned to use. They would use inactivated Sendai virus to fuse the egg cell, whose nucleus had been removed, with a cell from a mouse embryo. Then they would allow the nucleus of the mouse embryo cell to float into the egg and take over.

They began simply, asking: Could they transplant a nucleus from a fertilized mouse egg into another egg whose nucleus had not been removed? It was not a test of whether an embryo cell could be cloned. After all, these fertilized eggs were still just single cells; they had not even gone through their first division.

Solter's idea was simply to see whether his method would allow him to transfer nuclei from one mouse cell to another.

The method worked beautifully. The cells took up their new nuclei, grew and divided and flourished when Solter and McGrath transferred them to surrogate mothers. Within a few months, the first mice were born.

Then Solter and McGrath took the next step: transferring nuclei from a more advanced mouse embryo cell to a mouse egg whose nucleus had been removed. But it simply did not work. If they transferred a nucleus from a two-cell embryo, the egg would divide a few times, then die. If they tried transferring nuclei from older four-cell embryos or eight-cell embryos or the larger, forty-cell embryos that Illmensee and Hoppe had used, the eggs died right away, failing to develop at all. Since Solter and McGrath knew that their method of nuclear transfer worked, they could only conclude that the problem was not with the method of moving nuclei but with the concept of cloning, even cloning with embryo cells.

At the same time, Solter and McGrath tried to make mice with no father and mice with no mother. It proved to be biologically impossible, they discovered, because mothers, even human mothers, add proteins to their DNA in particular patterns, leaving some genes open and others blocked. Fathers pattern their DNA in a different way. And embryos can develop only if they have DNA with *both* maternal and paternal patterns. The phenomenon, dubbed "imprinting," spawned a new field of scientific inquiry.

Solter, however, had no idea what to do with the results of his experiments that had convinced him that Illmensee and Hoppe's findings could not be repeated. Scientific journals publish reports that investigators have done something, gotten a result, not that they have failed.

"It is very difficult in science to publish a negative result," Solter said. Why would it mean anything to readers to hear that someone was unable to do an experiment? All it might mean was that the scientist was inept.

"All we had were two very large sets of negative results," Solter told me. And Illmensee and Hoppe's positive results had been published and had been accepted as true. "I didn't have the slightest idea of how to persuade people of what we had," Solter said.

But Solter had repeated the experiments so many times that he was certain his results were correct. He sent a paper to the journal *Cell* saying that it is biologically impossible to create mice with just one parent. *Cell* was the journal that had published Illmensee and Hoppe's original claim to have cloned three mice. He also sent a paper to *Science* saying that he could not repeat the cloning experiment. Both papers were accepted. His paper in *Cell* appeared in May 1984 and his paper in *Science* in December 1984.

But it was the last phrase of the last line of Solter's paper in *Science* that reverberated through the halls of academe. He wrote, "The cloning of mammals, by simple nuclear transfer, is biologically impossible."

It was the worst of all possible outcomes. Was Illmensee's cloning work a fraud or was it a fluke? Could it be repeated if someone tried hard enough and long enough or was it truly impossible to clone a mammal? Who was right? Solter or Illmensee?

It's not that Illmensee had no defenders. Some, like Robert McKinnell, the cloning pioneer from the University of Minnesota who had worked on the cloning of frogs, says that the man was investigated and never charged with fraud. That was good enough for him.

"Karl went through hell," McKinnell said. And the thing that hurts the most, he added, was that his accusers were his own lab workers. The system reminds him of his Navy days, McKinnell told me. "In the Navy, if you ever were accused, you got a court-martial. And even if you were exonerated, it was a blotch on your career and you would never advance," he said.

Keith Campbell, Wilmut's collaborator, told me that he suspects that Illmensee really did clone those mice. With the cloning that led to Dolly, it is now clear that mammals can be cloned, even adult mammals. Mice, for unknown reasons, seem to be much harder to clone than sheep or cows or other large animals. But, Campbell said, he thinks that Illmensee got lucky, then could never repeat his experiment. That is when he got into trouble.

Nature magazine, which had published the most guarded short news stories on the Illmensee affair, nonetheless got angry letters accusing its editors of assuming Illmensee was guilty. Eckhard Lieb, a scientist at the University of Bayreuth in Germany, who said that he had known Illmensee for more than sixteen years and had worked in the same room with one of Illmensee's doctoral students, wrote that he was certain that Illmensee had not cheated in his experiments. "I am fully convinced of his personal and scientific integrity. If others fail to repeat his work successfully— what of it? Even Carl Lewis has not yet reached Bob Bemon's 8.90 metres." David Yaffee, a cell biologist at the Weizmann Institute of Science in Rehovot, Israel, wrote that he was "surprised to witness the relative ease with which many scientists accepted the truth of the charges (or even rumors before these publications) against a colleague, without knowing the relevant facts." He added, "A correct attitude towards scientific honesty should make it all the more difficult to accept that a well-known scientist has resorted to fraud. The instinctive reaction to such a charge (as long as proof has not been found) should have been disbelief."

Hans J. Becker, a biology professor at the Vienna University Medical School, sent a letter to one hundred scientists around the world, asking them to rally around Illmensee and write to the president of the University of Geneva on Illmensee's behalf. "I should start by saying that I felt shocked and sickened right at the beginning, that a series of offensive procedures on the part of members of Illmensee's group were able to throw doubt on the sincerity of Illmensee's work and thereby initiate such a grave series of insinuations and accusations, leading to the investigation of an international committee that has now presented its report," he wrote. And, he added, with a swipe at Solter, "it sickens me to hear—and if only as a rumor—of someone claiming to have repeated one of his experiments and to have failed to get his results; thereby—intentionally or not—insinuating another case of falsification."

Solter said he also got several letters addressed to him personally from outraged scientists, asking how he dared question Illmensee's work, especially in prestigious journals like *Cell* and *Science*. "I got letters saying I should be ashamed of myself. A lot of people in Germany believe he was pushed out of the University of Geneva because the Swiss do not like the Germans and that he was completely and unjustifiably hounded out of this work." But, Solter said, "you won't find any developmental biologist who believes that."

Except, perhaps, Peter Hoppe, Illmensee's collaborator. Hoppe knew that the only way he could remove the cloud of suspicion that was hanging over his head and over the head of Illmensee would be to repeat the experiments he had done with Illmensee. It was not enough that the committee investigating Hoppe had come up with the Scotch verdict: not proven. After all, as Solter notes, "What does it mean to say they could not find any evidence of fraud? Does that mean that you couldn't find any evidence and therefore believe there wasn't any or that you believe there was fraud but you couldn't prove it?"

So Hoppe felt, with some justification, that his future as a scientist was threatened. He became obsessed with the study that showed mice could be created with just one parent. The reason it had worked when he and Illmensee did it, he decided, was that they had used an especially good lot of a chemical, called cytochalasin B, which enabled the mouse eggs to begin to grow. Hoppe no longer had that lot of the chemical and when he tried to repeat the experiment with the cytochalasin B he had on hand, the experiment failed. His only hope, Hoppe decided, would be to find another batch from that lot or to find another lot that was just as good.

Solter was on a committee of the National Institutes of Health that reviewed scientists' applications for grants to do their research, and he vividly remembers a grant that Hoppe submitted at this time. "He wanted to test all the cytochalasin in the world and find one that worked," Solter told me. Hoppe did not get funding for this desperate attempt to redeem himself. He began to have serious personal problems. "His career went down the drain," Solter said.

Hoppe told me that he still thinks the experiments that he and Illmensee did together were correct. Illmensee did all the microsurgeries, Hoppe said, and was so careful that he refused to drink his morning coffee before operating on the mouse eggs and mouse embryos. "You are going to tell me that this was a guy who committed fraud?" Hoppe said. "He suffered for three hours without a cup of coffee. If he was going to commit fraud, he'd have gone ahead and had his cup of coffee."

And what about the fact that Illmensee could clone only embryo cells, not more advanced cells? Hoppe asked. "If he was going to commit fraud, why not do that? It would have been fantastic."

No, Hoppe said, the problems occurred only after Illmensee and he stopped collaborating. Although he and others were unable to repeat the spectacular experiments that he and Illmensee did together, he said that might be because few had Illmensee's patience and his technical skill and because people just did not try hard enough with exactly the same methods that Illmensee had used.

Many scientists think Hoppe was bedazzled by Illmensee and had no idea that the experiments would be unrepeatable. In any event, they say, his suffering was so enormous that they can feel only the deepest regrets and wince in pain as they recall the way Hoppe's career went down in flames. Hoppe's reaction to his association with Illmensee "essentially destroyed not only his career but also his personal life," Tilghman said.

No one else spent much time trying to make Illmensee and Hoppe's experiments work. Solter's collaborator, James McGrath, went to medical school and is now a geneticist at Yale, no longer working on mice or cloning. Solter went to Germany, fed up with the American way of doing science, and focused his research on other matters. His flight to Germany, said Brigid Hogan, the developmental biologist at Vanderbilt University, was "a tragedy for American science."

Solter sees it differently. "For twenty years, I was living on nothing but grants, including my own salary," Solter said. "I had a lab with thirty people and I had two million in grants. I found myself doing absolutely nothing but writing grant applications, writing applications for grant renewals, writing reports on grants, and so on. I feel this was not the way to live."

In addition, Solter said, in the United States, labs are supposed to grow. "When you stop growing, people have a tendency to think you are ready to die. But as your lab gets bigger, you get less and less efficient and you stop spending time with students who are not extremely productive. I felt I was abandoning them. I felt I should be spending more time with people who are not extremely productive."

And finally, he felt that with the growing pressure to have a huge lab and to jump on problems as fast as you can to beat the competition, there was a certain loss of reflectiveness, of creativity. "I'm now fifty-five," he told me in 1997. "I grew up in a time when you could say almost nothing was possible" because there were no tools and few special instruments and exquisitely sensitive biochemicals available through glossy mail-order catalogs to allow molecular biologists to do their wizardry. And with no tools, Solter said, scientists spent all their time thinking of how to plan a study so it would be meaningful and so that, whatever the results, they could learn something.

And so, when he was offered a position as head of a lab at the Max-Planck Institute in Freiberg, Solter saw a chance to do science in a way that seemed to be lost in America. He was given a budget to do with as he willed. So he set up six independent groups to work intensively on problems that interested them. His own lab, he said, has just four people. He writes no grant proposals.

"A lot of people would say that science today cannot be done that way, and sometimes I agree." But, Solter said, at least he has to think carefully about what he wants to do and why he wants to do it rather than rushing to beat a dozen other labs to do the next obvious experiment.

Solter said he lost interest in cloning for three reasons. First was the discovery of imprinting, which seemed so much more rewarding to pursue because it was a real, repeatable phenomenon and it showed the astonishing fact that genes inherited from the mother are not the same as genes inherited from the father. It was a serious challenge to the textbook laws of inheritance that were first discovered by the monk Gregor Mendel.

Second was the abysmal success rate with cloning attempts. The mouse researchers did not need to clone a mouse to study developmental biology. All they wanted was to be able to transfer the nucleus from a mouse embryo cell into a mouse egg and have the egg live long enough for them to study it. Virtually none of the eggs, however, lived even a few hours with their transplanted nuclei. If scientists wanted to ask the fundamental question, Why is the nucleus of an early embryo cell different from that of a cell that is more advanced in development, they would have to be able to examine dozens of cloned embryos, even if those embryos lived for only a few divisions. "We said, 'Forget it. We can't do anything with this success rate,'" Solter told me.

Finally, Solter lost interest in cloning because he realized that scientists knew almost nothing about which genes were active in early development. If he wanted to transfer nuclei to study how development is controlled, it would be best if he knew ahead of time which genes were supposed to be turned on in a newly fertilized egg and in the first few divisions of an embryo. Those genes would serve as molecular markers for development—if they were working it would mean that development was proceding normally. If he knew those genes, he could ask whether the transferred nucleus turns on those same crucial genes and, if not, why.

By the mid-1980s Solter had completely shifted his interest to a search for these genes that are needed early in development. It was a huge project and pushed cloning out of his mind entirely.

Illmensee told me that he and Hoppe never retracted their disputed papers because there was no reason to retract them—they were correct. But, Solter said, "it didn't make any difference. We went from a couple of years when everybody believed them to a time when nobody believed them. It became totally irrelevant to go and investigate whether Illmensee had committed fraud when nobody believed his results anyway. Everyone said, What difference does it make? His career is anyhow ruined. Isn't that enough?"

Solter said he was not discouraged by his result, and his conclusion that cloning of mammals was impossible. In fact, he thought that it was much more interesting if cloning could not be done than if it could. For scientists, he said, it opened up much meatier problems. If genetic material irreversibly changed as soon as an egg was fertilized, there were all sorts of questions to investigate. What precipitated the changes? What were the changes? "I always thought it was more interesting if you had problems with reprogramming," genetic material. "That means there is something to investigate."

Other mouse researchers, said Lee Silver, a Princeton University professor, moved on to other questions. It was so much more interesting to unravel the mysteries of imprinting. It was so much more interesting to study mouse genes and to create genetically engineered mice. Anyway, who would pay for cloning research? Who cared about it anymore? There were other ways to study development.

But others were devastated by Solter's paper. It was not just Solter's fame, not just his credibility that stunned them. It was his strongly stated conclusion that not only could he not repeat the experiment of Karl Illmensee and Peter Hoppe but that cloning of mammals was actually "biologically impossible." In a world where scientific research depended almost entirely on getting grants, and getting grants depended on convincing someone that not only were you capable and well equipped but your project was achievable, a statement like Solter and McGrath's was the sounding of a death knell for the future funding of cloning.

Those in traditional cell biology and molecular biology labs who wanted to continue cloning studies found themselves scrambling for funds. Bob McKinnell had always been supported by the National Institutes of Health, by the American Cancer Society, by the National Science Foundation. His work involved nuclear transfers into frog cells, examining how easily cancer cells could be reverted to normal if they were cloned. But times had changed.

"Cloning was just bypassed, just left behind," McKinnell said. "I wrote my research proposals and they came back, turned down." It had never happened to him before, and now it was happening all the time. "Some things just are not trendy," McKinnell said.

Finally, he found a sponsor. It was the Council for Tobacco Research, which supports basic research related to cancer. "They were great people. And they were also sweet, so sweet," McKinnell said, ready to oblige his every request.

He defiantly announces their support. "At every talk I give, I flash on the screen a slide that says, 'Funded by the Council for Tobacco Research.' People look at me like I've got the plague or something."

Illmensee, in the meantime, dedicated his life to clearing his name. He told me that he repeated the unpublished experiments that had been questioned by his lab workers and published the results. He faxed me a copy of a letter to the University of Geneva from two British members of the international committee that investigated him, Ann McLaren and R. L. Gardner, noting that "the essential findings of the earlier papers have now been reproduced. The two journals in which the papers were published (*Development, Naturwissenschaften*) are both internationally recognized peer-reviewed journals." And so, they concluded, "As members of the International Commission we consider it appropriate that the University of Geneva inform the scientific community that these controversial experimental findings have now been confirmed under conditions specified by the Commission."

Since 1991, Illmensee told me, he has been writing letters to the University of Geneva asking for an apology and a statement that his experiments had not been fraudulent. "I have never received a single letter from the university during all these years," Illmensee said. "A lot of my colleagues have also sent letters to the university and they have never received an answer. This is for me a very unfortunate situation. My suspicion is that the university wants to cover it up," he added.

The disputed experiments, of course, were not the ones in which Illmensee said he cloned mice, but Illmensee said they tainted his reputation so much that the cloning experiments fell under suspicion as well. "There was an opinion and a strong group that could not show positive results," Illmensee said, obliquely referring to Solter and his supporters. "There was a sort of power struggle and I was at that time on the weak side," he added. Now, with the birth of Dolly, Illmensee told me that he feels vindicated. Solter was wrong that cloning was impossible and Solter was wrong that Illmensee's experiments were not repeatable.

But Illmensee was correct that with the publication of Solter's paper, he lost the power struggle. When the cloning claims began to seem dubious, the reaction of many scientific leaders was extreme—not only did they turn their backs on the man who said he had cloned but they began to disdain the very pursuit of cloning itself. Their reactions sent a message to those who were deciding what research projects to pursue and, just as important, to those who decide which research should be funded. Scientists whose votes in committee meetings at the National Institutes of Health determined how millions of dollars in federal research money would be spent became uninterested in grant proposals on cloning. Animal scientists who wanted to clone sheep and cows for the most prosaic reasons found themselves desperately defending work that just a few years earlier had been deemed exciting and promising.

Everyone who cared about cloning, from Robert McKinnell, the cloning pioneer who worked on frogs, to Ian Wilmut and his colleague Keith Campbell, had to take a stand—did they accept the mouse work or not? Did they believe a charismatic German researcher who insisted he had cloned mice or did they believe the sober paper in *Science* magazine by a Croatian emigré that said mice could not be cloned? And if they believed that mice could not be cloned, did they believe that meant that cloning in general was a biological impossibility in mammals? The rise and fall of the cloned mice was an event that still echoes in the halls of science as researchers ask themselves how that strange tale could have developed the way it did.

Now, with the birth of Dolly, the cloned mice have become an issue again. Were those scientists who eventually rejected the work wrong? Could it be that careers were ruined and cloning research was marginalized for naught? Was the saga of the cloned mice a comedy of errors or a Greek tragedy? Wherever the truth lies, the impact of those three mice on cloning continues to resound.

With a few exceptions, like McKinnell's work, cloning left the high-profile world of molecular biology and retreated to the little-known world of animal science. This, of course, explains the enormous surprise of leading scientists when they learned that Ian Wilmut and Keith Campbell had made a clone named Dolly. Who *were* these scientists? some molecular biologists plaintively asked. The answer, of course, is that they were the animal scientists, the only ones brave enough to take up the cloning problems. They were a community unto themselves, unknown to the elite circuit players of molecular biology. And they had been motivated more by the economic promise of cloning than by a thirst to understand the molecular mysteries of development.

7

BREAKING THE LAWS OF NATURE

The role of the scientist is to break the laws of nature.
—STEEN WILLADSEN

Conventional wisdom says that research flows from university laboratories, where scientists follow their interests for the sheer joy of making discoveries, to company labs, where the focus is on the bottom line. The notion is that the brilliant ideas, and the insights that require intellectual freedom and unfettered imagination, come from work that is funded by government agencies like the National Science Foundation or the National Institutes of Health or by private groups like the American Cancer Society. Companies and agencies interested in pragmatic research come in afterward, exploiting the fruits of basic science.

But cloning was an exception. At a time when the vast majority of scientists felt that it was useless—and uninteresting—to spend any more time attempting to clone, an intrepid handful of researchers who were supported by business concerns soldiered on, their work ignored by most molecular biologists.

These cloners worked in agriculture departments, surrounded by the sounds and odors of barnyard animals. Their work had utterly pragmatic aims and their publications, even in major journals like *Science* and *Nature,* went mostly unread by the glitzy research community that had grown up on genetic engineering, that worked at a feverish pace on what developmental biologist Davor Solter, the man who declared that mammals could not be cloned, would call the next obvious experiment. The animal scientists were not part of the elite group of developmental biologists who gossiped about each other's research at meetings like the Gordon Conferences. The circuit that attracted the molecular biology and molecular genetics leaders did not include people who studied large farm animals like sheep, cows, and pigs.

"These agriculture people, whom I've never met, don't talk to the mouse developmental biology people," said Lee Silver, the director of a molecular biology lab at Princeton University. "We don't ever come to the same meetings."

Since researchers like Silver are using mice to study human diseases and human genes, their work is noticed by the media, by the public, by the National Institutes of Health. The scientists working in barnyards and gathering eggs from ovaries that they collected at slaughterhouses were increasingly isolated from the science superstars, whose universe seemed to be the only one that mattered.

The saga of the scientists who proved Solter wrong is a tale of two labs, one in the United States and one in England. The American lab, at the University of Wisconsin, was run by a senior professor and funded by a giant corporation, who assigned a young graduate student to the cloning task. The British lab was the domain of a Danish iconoclast who was funded from year to year by the British Milk Marketing Board. He worked in an agricultural research station where there was no pressure to publish and where, he said, the senior, tenured scientists had what amounted to a sinecure and were free to do nothing if they so pleased.

Most scientists would recognize the workings of the American lab. Although it dealt with cows, not mice, and although its goals were less to understand the mysteries of development than to produce cloned cows for industry, its manners and mores were familiar. The only true surprise was that its researchers were so amazingly successful in the face of enormous inexperience with the new techniques needed for cloning.

The British lab was a strange place that would seem totally alien to most scientists. But the man who was working there on cloning was one of the most unusual and brilliant scientists the twentieth century has seen.

At the University of Wisconsin, the news that Karl Illmensee's work in which he claimed to have cloned mice from early embryo cells was not repeatable came as a staggering blow. The Wisconsin animal-science lab was directed by Neal First, an ebullient optimist who was not only ready to clone, but who also had funding from W. R. Grace and Company that assumed he was going to clone. The idea was to use cloning to multiply valuable cattle embryos. First was supposed to figure out a way to take each cell of an eight- or sixteen-cell embryo and clone it—use it to create a new animal that would be the identical twin of the animal the embryo would have become, were it allowed to grow. To clone an eight-cell embryo, he aimed to take eight eggs whose nuclei were removed. He would add to each egg the nucleus of one of the embryo cells. If the process worked perfectly, each egg would grow into an embryo that was a clone of the original eight-celled one. Then he hoped to repeat the process, using each eight-cell embryo as a source of eight new cells for cloning, creating sixty-four identical embryos. He thought they might be able to go on again, using each of the sixty-four embryos as a source of cells for cloning, making 512 embryos, then do it once again, creating 4,096 embryos, and so on. "We wanted to create an unlimited source of identical animals," said Randall Prather, who was a doctoral student in the lab at the time. Since embryos from the best cattle can cost $500 to $1,500 apiece, being able to clone multiple embryos seemed like a real money maker.

First told me that his group had initiated the entire cloning project shortly after Karl Illmensee and Peter Hoppe published their paper in 1981 saying they had cloned three mice and that they had based their entire project on that paper. First's cloning team included a postdoctoral student, James Robl, and eight doctoral students. Working in a tall brick tower overlooking the stock pavilion at the university, where agriculture students exhibited their prize animals, the scientists were determined to make cloning work. First began by inviting Peter Hoppe to come to the lab and demonstrate how to move the nucleus from an embryo cell into an egg. After Hoppe's visit, First asked Robl to learn the method. Robl set to work, using mouse cells. But the cells kept dying, and Robl and First could not understand why.

Then came that fateful paper by Solter and his postdoctoral student, James McGrath, that concluded that the cloning of mammals was biologically impossible.

"Man, it was depressing," Robl said. "I remember Neal flying out to New York the next day to talk to Mr. Grace himself about why we are spending all this money, to justify why we are spending money on cloning." And Robl remembers that a famous scientist, Thomas Wagner, came through the lab. "I showed him with enthusiasm all the work I was doing. He looked at me with a very serious look on his face and said, 'Why are you doing this?' I felt like I was about two inches tall," Robl said.

Even First was disheartened. He said he does not remember that journey to talk to Grace, but concedes that Robl remembers several things that he may have pushed out of his mind. But he cannot forget how he felt when he read that *Science* paper. McGrath and Solter's conclusion "set things back a long way," he said. "It said to anybody trying to do a cloning experiment that nobody approves of you very much and the experiment probably isn't going to work."

Still, the scientists in First's lab had no choice but to regroup. After all, if they decided they could not clone, they would have no money from W. R. Grace. And if they had no money from W. R. Grace, there would be no lab, or at least no project and no funding for Robl, Prather, and many of the others. So, First said, they told themselves that maybe it was still possible to clone. Where was it written, after all, that what was true for the mouse was true for the cow? "We said, 'There are ways around this. After all, this was the mouse.'"

Robl soon left to go to the University of Massachusetts as an assistant professor, and Prather inherited the project to actually clone a cow. He was excited but wary, yet he was not really in a position to refuse. He wanted, of course, to share the glory of achieving the seemingly impossible. But he also needed to protect himself from abject failure at a time when all he really wanted was to get his degree. He knew all too well that doctoral students like himself have to be careful when they select their research projects. If they choose something too hard, or if their adviser, who is a mentor, is too remote, looking on from a distance and not really taking a personal interest, or is not helpful, they can fall by the wayside, years of training gone for naught. And they know how easily that can happen, because even in the best of schools, where getting admitted to a Ph.D. program is as highly competitive as getting into an elite college as a freshman, many, if not most, students who enter the program leave without a degree. Unlike medical schools, where those who enter as first-year students are pretty much guaranteed that they will graduate, Ph.D. programs offer no certainty and a lot of anxiety.

The only way to get a Ph.D. degree is to do original research—to discover something that no one else has discovered before. But you don't have to break entirely new ground. The goal is to get out, degree in hand, get a job, get some grant money, and then try to make your name.

So Prather thought he saw the handwriting on the wall. Cloning could be the end of him as a Ph.D. student—it was the impossible project, the unproved ground. To make matters worse, he had a plan for his doctoral research, and he felt it was unfair for First to require him to embark on such a risky project. Prather's partner in the work was Frank Barnes, another doctoral student, whose task was to figure out how to grow the calf embryos after Prather cloned them. Prather thought it was just too much responsibility. "The funding for much of the lab hinged on Frank and me making it work," he said. Here he was, just out of college, just out of his teens, in fact, and he was supposed to clone and prove that Davor Solter, a senior scientist with a towering reputation, was wrong. The only way he could dare to do it was to hedge his bets and pursue two projects at once—the cloning work and a study of mouse embryos that involved mixing cells of different embryos together in order to create mice with a mixture of genetic traits, a project that he actually used to get his doctoral degree.

It was one of those ironies of fate that Prather was in Neal First's lab at all. He had only wanted to be a Kansas farmer, to take care of cattle. Prather grew up in Manawa, Wisconsin, about fifty miles from Green Bay, the son of a veterinarian whose special interest was dairy cows. When he was fifteen, his parents took a chance on their dream, sinking their life savings into a beef and grain farm in Kansas and buying three hundred cows. Prather went to Kansas State University, but he did not pay much attention to his grades because what did it matter? He was just going to return home and be a farmer with his parents. Finally, in his junior year of college, he became captivated by embryology, fascinated by a course in reproductive physiology. At the same time, he said, "the farm didn't look so good." His parents' business was failing and he felt that perhaps the best thing for him to do would be to get a master's degree and then, when the business improved, return home to farm.

But after Prather got his master's degree, life on the farm still looked grim. The farm was in east central Kansas, a place where the weather was unpredictable and droughts were common. For Prather's parents, a combination of dry years, high interest rates, low grain prices, and declining land values spelled disaster. Finally, in the mid-1980s, Prather said, his parents gave up and his father returned to Wisconsin to be a vet once again.

"I decided, well, I'm not going to be able to go back to the farm, so I guess I have to get a Ph.D.," Prather said. He thought that he would get the degree and settle into an academic position.

He ended up at Neal First's lab at a time of great anguish. Solter's paper had just been published, and the lab was trying bravely to regroup. Prather found himself in the midst of the upheaval. With First's prompting, he finally decided to take Solter's claims as a challenge. "We were going to make it work," Prather told me. Moreover, he had heard a rumor that the formidable Danish scientist Steen Willadsen, working in England, was doing the same thing, and that just fueled Prather's competitive fire.

But the task ahead was daunting. Cloning cattle was nothing like attempting to clone mice. With mice, embryo cells grew well in the laboratory, in small plates of nutritious broth. When they had grown for a few days and were ready to implant themselves in the uteruses of surrogate mothers, it was a simple matter just to inject them.

Cows were a different matter. Nothing was simple when it came to working with these huge animals whose embryos did not take kindly to laboratory conditions.

The first step was to find a way to grow cattle embryos in the laboratory for a week until they became a ball of cells known as a blastocyst, the stage when they could be transferred to the womb of a cow. The scientists would go to the local slaughterhouse, bring back some cow ovaries, pluck the speck-sized ripened eggs out of them, then fertilize the eggs with semen from bulls. Finally, they tried to coax those fertilized eggs to divide and grow.

When the cloning project in Neal First's lab began, only 5 percent of fertilized cow eggs survived long enough to grow to blastocysts. There would be no point in cloning cells if the embryos died of their own accord, even without new nuclei that would be added in cloning attempts. One of the other doctoral students in the lab, Willard Eyestone, was responsible for figuring out how to keep the embryos alive.

"You are confronted with two options," Eyestone said. "One is to take a freshly fused egg and leave it out of the cow only as long as you absolutely have to. But imagine this scenario. You have to put the cow on her back and surgically go in and get access to the oviduct, transfer the embryo into this oviduct, then put the animal together, knock on wood, and all the rest. That would be one way to do it." It is a method that is not too burdensome if you are working with mice or even sheep, Eyestone told me. But cows—that's a different matter.

If scientists can keep an embryo alive for a week outside the cow, then there is a relatively easy way to put it into the cow's uterus, where it will grow. They inject the embryo, which is about the size of a grain of salt, through a stainless steel syringe, threading the nearly two-foot-long syringe through the cow's vagina, through her cervix, and into her uterus. "Judging from the cow's reaction, this is nearly pain-free and at most mildly irritating," Eyestone said.

So Eyestone decided to find a way to grow cow embryos for a week outside of a cow. He saw a paper by Steen Willadsen reporting that Willadsen had started with eight-cell cow embryos and had grown them successfully in sheep oviducts. The method still required operating on a sheep and inserting the embryos in the oviducts, sewing the animal up again, then coming back a week later, surgically removing the oviducts, taking them back to the lab, and flushing out the embryos. But it was much easier to operate on a sheep than on a cow. For one thing, sheep were much smaller.

If the method worked for eight-cell cow embryos, "we reasoned it wouldn't be much of a leap to start with a one-cell embryo," Eyestone said. Why struggle to keep a newly fertilized egg, a one-cell embryo, alive in the lab until it grew to eight cells? Why not just let it incubate in a sheep oviduct? "And *voilà,* we had a means now by which we could take freshly fertilized embryos, put them in a sheep oviduct, culture them to the blastocyst stage, and transfer them to a cow.

"We were initially thrilled that we could do this. But the neatness of it wore off fairly quickly. It was still a lot of work," Eyestone told me. He reasoned that perhaps he could simply remove the oviducts from sheep and keep them alive in the lab, adding the cow embryos to the oviducts without having to keep operating on a sheep. The method succeeded. The oviducts could live in the lab for about a month, providing nutritive substances for the embryos. And what were those special chemicals secreted by the oviducts? "We never did find out," Eyestone said. "To this day, we don't know."

With the problem of growing the cow embryos solved, it was time to clone. Prather optimistically decided that there was a good chance cloning would work in cattle, even though he was sure that Solter was right that cloning could not work in mice. "There are so many differences between cattle and mice," Prather said, that he decided it would be folly to treat the two species as if they were the same. "It was almost a crusade," Prather told me. "And if it went against people's feelings, well, that's just because all the research so far had been done in mice. We didn't know any better. We knew mice were different from cattle and it was almost an insult to us that people assumed that all domestic animals were the same as mice." It was, Prather said, "a kind of defiance," one born of desperation.

Meanwhile, Prather and the others in First's lab kept hearing rumors that Willadsen was doing the same thing in England.

They tried to find out what Willadsen was up to, using Neal First's contacts through a circuit of animal husbandry researchers that he was on, but they got only the sketchiest details.

Somehow, though, their own experiments were succeeding. They had discovered that they could use a machine that was sold to fuse cells for other types of experiments to fuse an embryo cell with an enucleated cow's egg. The machine used a brief burst of electricity to meld the two cells together.

"All of a sudden, we were able to make a few things work," Prather said. They might not have cloned a cow yet, but they could at least fuse a cow embryo cell with a cow egg, making a new embryo that was a clone, and get the clone to grow to eight-cell stage.

Prather prepared to make his announcement. A major scientific meeting was coming up, and he submitted an abstract. He was scheduled for a poster session. Almost all scientific meetings have these sessions in which scientists stand before poster boards on which they have tacked a description of some experiment they've done. Others at the meeting stroll from poster to poster, reading the ones that are most intriguing, questioning the eager scientists standing sentry next to their posters. It was a step below a talk before an audience, but significant for Prather nonetheless.

When W. R. Grace administrators heard of Prather's plans to announce his results in a poster session, however, they forbade him to do so. Although, Prather said, the abstract did not reveal anything confidential, the company was wary. "They said, 'You can't present. We don't have patent coverage.'" In the United States, companies have one year after work is publicized in which to apply for a patent, but if companies want world patent rights, they must apply for the patent before the work is published. W. R. Grace wanted world rights.

"So I went to the meeting and there was a blank space where the poster was supposed to be," Prather said. "It was kind of awkward." Scientists asked him why he was at the meeting and he had to tell them that he had some results but he wasn't allowed to present them.

In the fall of 1986, Prather, Eyestone, and other students in the lab finally succeeded completely—they transferred the nucleus from an early cow embryo to a cow egg, grew the embryo in a sheep's oviduct until it reached the blastocyst stage, and transferred the embryo to a surrogate mother cow. Ten months later, after the normal cow gestation time, the animal gave birth to a calf that was a clone of the embryo they had started with.

But Prather was not the first to clone a farm animal from an early embryo cell. In March 1986, Steen Willadsen announced in a paper published in the prestigious British journal *Nature* that he had cloned sheep from early embryos.

Prather and First had their paper on the cloned cows ready to go by August 1987. Now, the question was, where would they publish it? "I told Neal, 'I want this published in 1987. If we wait until 1988, it will look like we did it after Steen and it took us two years to get it published. We have to have a 1987 date on it."

At the time, Donald Dierschke was editor of the journal *Biology and Reproduction,* and he was a professor at the University of Wisconsin, just across the campus from Neal First's lab. Prather begged Neal to send the paper to Dierschke and to persuade him to publish it in 1987.

First called Dierschke about the paper, telling him that he would give the paper to *Biology and Reproduction,* but only if Dierschke would expedite it and publish it that year. Otherwise, First said, they would send the paper to *Science.*

Dierschke demurred, saying he was not sure he could do that. First reminded him, "You're the editor, you can do whatever you want." It came out in November.

Prather, Eyestone, and a handful of other young doctoral students had done it—they had completed the impossible cow-cloning project in Neal First's lab. And they knew they had proved once and for all that cattle were not gigantic mice. It might be infeasible to clone a mouse, but it certainly was feasible to clone a cow, at least from embryo cells. Yet no one in that lab, certainly not Randy Prather, certainly not Neal First, was thinking about cloning from cells from older embryos. They definitely were not considering cloning from adults, as Ian Wilmut did when he cloned Dolly. "We thought it was impossible," First said.

Steen Willadsen, at work at the British Agricultural Research Council's Unit on Reproductive Physiology and Biochemistry in Cambridge, England, had no such reservations. He did not accept the notion that there should be insurmountable obstacles to cloning adults. In fact, throughout his scientific life, he had scoffed at the very idea that hypothetical biological or technical barriers might stymie him. "The role of the scientist is to break the laws of nature, rather than to establish, let alone accept them," he told me.

The near-mythical Willadsen has played the role of the brilliant, secretive innovator ever since he began working with animal embryos in the 1970s. Although Davor Solter did not know Willadsen—they came from different scientific worlds—within a year after Solter had published his paper saying that the cloning of mammals was impossible, Willadsen's first cloned sheep were being born. Granted, the sheep were cloned from early embryo cells. But that, after all, was what Solter had said was biologically impossible. Years later, when Ian Wilmut and Keith Campbell decided to try to clone an adult sheep, Wilmut said he got his inspiration to try from rumors of an experiment Willadsen had done—but never published. Wilmut said that while he worked on the cloning of Dolly, he suspected that he might have just one competitor in the world who could do it before him—Steen Willadsen.

Of all the scientists who have cloned, or tried to clone, Steen Willadsen is perhaps the most grandiose and the most visionary.

He also is the most mysterious, a man who seemed to spring up out of nowhere, ready to do the impossible. In actuality, Willadsen developed his golden hands and his clever methods that made cloning work through determination, keen intelligence, single-minded focus, and incredibly hard work.

Willadsen eschews the ordinary workaday world of the academic scientist. What he fears most is what he calls the dead-pheasant syndrome—"though still airborne with wings flapping, the bird is already dead," he explained. It is, he said, an occupational hazard of scientists, who can keep publishing and appear active long after they have lost their zeal. Willadsen's aim was different. It was, he said, "not just to become a scientist, but to engage in great, absorbing endeavors while maintaining a high degree of freedom, and avoiding tedium and coercion."

In 1997, when I visited him, Willadsen was fifty-three, but he looked much younger. He was slim and casual, with a shock of light brown hair flopping over his forehead. He drives quickly in an old Honda, tosses off judgmental statements ("Of course you don't like Disney. They are fascist") and tends to outbursts that he immediately confesses sound brash, or arrogant. And he loves to talk. Every question elicited a long answer, often far more interesting than might be anticipated. And every time it seemed that Willadsen was finishing his story, he would roar out "BUT" and continue on.

Willadsen lives in a huge pink-stucco house in a new development in Windermere, Florida, near Orlando, with thirteen-foot ceilings and a swimming pool on an enclosed porch. He denied that his house was large or terribly luxurious, although, he noted in one of his inadvertently brash asides, "it's not for the underprivileged." Anyway, he said, it's actually his wife's house. He spends his days tending his lawn, writing, and working part-time at two in vitro fertilization centers, one near his home in Florida where his wife, Carole, a vivacious embryologist who speaks rapidly in the brogue of her native Scotland, is the scientific director, and another in New Jersey. In between helping infertile women have babies, Willadsen experiments with mice and with human eggs that would otherwise have been discarded.

Willadsen was born in Copenhagen but spent most of his childhood on a dairy farm in Jutland, Denmark, where he lived with his divorced mother and her family. He rarely saw his father and had little to say about him. "What did he do? He wasted money. He was a spoiled brat of an earlier era. That's a simplified version of it," Willadsen said.

When Willadsen was growing up, the youngest of three children, he thought little about what he might want to do with his life. "I liked farming, in a way," he said. "But at the same time I could see that farming wouldn't be particularly great from a material point of view."

He went to Danish public schools, without worrying about his education. His family paid little attention to his schoolwork. Then again, no families in his town in those days paid much attention to their children's education. Willadsen said ruefully that he and his wife spend more time in one month overseeing the schoolwork of their two children than his family spent overseeing his work in his entire time in school.

But Willadsen had no trouble with academic work. "I was uniformly good at anything I wanted to be good at. By the time I finished school, I could do whatever I wanted to do. That was the reality," Willadsen told me.

After high school, Willadsen decided to go to veterinary school (in Denmark, students go directly from high school to vet school) at the Royal Veterinary College of Copenhagen. "It was not a hard decision. I drifted into it," he said. He and his older brother Claus, who also was a veterinarian, had a vision of living a bucolic life as the local vets in a small farming town. But the reality of working as a veterinarian soon began to wear on Willadsen.

"I worked in various places—I was very keen to get a lot of experience—but then I decided that I was a little bit dumb. I worked for older vets, and although I wasn't in it for the money, it seemed silly that I could do everything that was required to run a practice—in fact, I was the one doing the work—but I was not running my own." He was twenty-seven at the time and looked as if he were in his late teens; he thought he looked too young to be accepted as the director of a veterinary practice. Besides, he was beginning to think that it would be boring.

The denouement came one day when Willadsen was assigned, once again, to work in the local slaughterhouse. In those days, veterinarians were responsible for all meat inspection in Danish slaughterhouses. "You were a somewhat glorified abattoir worker," Willadsen said.

That morning, Willadsen watched the pig carcasses come swinging down the line and looked over at the man working next to him. "His job was to remove the kidneys, liver, lungs, trachea, and tongue from the dead pigs, all together. It was like watching an alien, the way he tore them out. I could see that he was not a happy man," Willadsen said. "I decided then and there that I would not get into a routine like that."

Willadsen had made up his mind, and now the only question was what to do. He thought that just two subjects were "worth wasting time on"—the brain and reproductive physiology. He chose reproductive physiology, returning to the Royal Veterinary College in Copenhagen to study for a doctoral degree. It was a new graduate program, started just the year before, and it was housed in the veterinary school. "The vet school at that time was considered a place that did not belong in academe," Willadsen said. "It was considered a very bad place, and, on top of that, it had gained a bad reputation, due mainly to its internal politics, and several of its professors had been publicly branded liars and swindlers."

His Ph.D. project was to try to mature cow eggs in the laboratory. Cow eggs, like the eggs of other mammals, normally mature in the ovaries, ripening under the influence of hormones. "Nobody in the department had ever seen a live cow egg, or any mammalian egg, for that matter," Willadsen told me. His professor had found three research papers that seemed related to Willadsen's project, which he gave to Willadsen to read. Then he told Willadsen, "Go to the slaughterhouse and get some ovaries and take it from there."

Willadsen pulled it off. "I actually think it was not bad work, although it was poorly supervised," Willadsen said. "The people who had never seen a cow's egg before thought it was fabulous."

In the process of doing this project, Willadsen had learned something that would prove invaluable. "I became very good at handling eggs," he told me, explaining that it is a skill that many who attempted to clone had not bothered to acquire. It is a tactile feeling a scientist develops while working under a microscope, learning just how much to probe an egg without fatally injuring it. And it is a skill that can be learned only by practice. "The rest you could read about," Willadsen said.

When he finished his doctoral degree, Willadsen cast about for something to do with it. He saw nothing that interested him in Denmark, but at that time, the British Agricultural Research Council's Unit on Reproductive Physiology and Biochemistry was the mecca for livestock embryology, so he decided that that was where he would go.

Ian Wilmut, the man who, working with Keith Campbell, cloned Dolly, was there as a doctoral student, discovering how to freeze the sperm of boars. He left for the Roslin Institute just before Willadsen arrived and Willadsen inherited Wilmut's fellowship and his research project. Yet despite the remarkable research that was almost routine at the Cambridge center, there was no waiting list of scientists demanding to work there. "It was not exclusive because nobody was admitted. It was exclusive because hardly anyone wanted to go," Willadsen told me. Many years after he left the place, he asked Christopher Polge, the scientist who was his immediate supervisor—and who also had been Ian Wilmut's adviser—why he had been chosen. "He said there were no other applicants," Willadsen said.

At the British lab, the senior scientists were indifferent as to whether or not they or their students published papers, taking the attitude that publications were not the sine qua non of science. This suited Willadsen ideally since he was out to do great things that were so spectacular that they could be described in a few words; he had no interest in writing endless research papers reporting on technical minutiae.

Willadsen's favorite journals were *Nature* and an obscure publication called *The Veterinary Record*. He liked *Nature* because it was obviously prestigious—scientists would fight to get their papers published there. Equally important, he said, the editors were less petty than those of most other journals. If the work was stunning, they would tolerate a certain amount of sloppiness. "*Nature* wanted to publish papers with great ideas. It didn't matter if every last *i* was dotted or *t* crossed," Willadsen said. *The Veterinary Record* would publish anything of interest to veterinarians, with a minimum of editorial fuss, and so Willadsen liked to use it to publish practical, but less original, observations. "I didn't have to worry about picky comments from referees," he told me.

"I had a colleague who used to rank journals with stars. *Science, Nature,* and *Cell* in his estimation were five-star journals. *The Veterinary Record* got no stars. On one occasion he said to me, 'Why don't you just throw your paper directly in the garbage can?'" Willadsen said.

It was a difficult life at the Cambridge research center. Willadsen, as a Milk Board fellow, was supported year-to-year on a pittance: paid just £80 a month, or $150. Although he was single and had no particular interest in material possessions—"I could live on a stone," he told me—even he was thwarted by the salary. For the first and only time in his tenure there, he complained, and got a raise to £120 a month. "I never discussed promotions and I never discussed salaries. I thought it was their job to see to it that I got a reasonable salary and it was their job to see to it that I was promoted."

But although the pay was miserly, Willadsen was in perhaps the perfect place to discover how to clone. Unlike Neal First, who was fettered by his contract with W. R. Grace, and unlike mouse researchers, who were supported by grants from government agencies, Willadsen was a free agent. He had plenty of animals to experiment with, he had laboratory space, and he had utter liberty to do whatever he wanted, whether or not it had a chance of succeeding, whether or not it would ever result in a publication, and whether or not it advanced knowledge in the field. While he did have his eye on practical applications of his work, he was not forced to restrict himself to any one in particular. It was perhaps the combination of Willadsen's veterinary training, his talent for working with eggs, his curiosity and imagination, and his endless hours of work that was the magic potion that allowed cloning research to rebound.

Willadsen's first project, and the only one that was assigned to him during his time at the unit, was to perfect the freezing of livestock embryos, and particularly cattle embryos. He needed to make embryo freezing easy and efficient. Cloning, at that time, was not part of his or anyone else's thinking, but, of course, freezing embryos would turn out to be crucial to the cloning business. If he was going to be creating lots of embryos, it was crucial that he be able to store those he did not want to implant immediately in surrogate mothers. And that meant freezing the embryos.

At that time, there were three dictums about embryo freezing. One was that you had to replace the cells' water with dimethyl-sulfoxide (DMSO); no other chemical would do. Another was that you had to freeze the cells very slowly. The third was that you had to thaw the cells equally slowly.

"I made it my job to disprove all three," Willadsen told me. "They were too narrow and dogmatic and they set embryos apart from other cells." He succeeded. Yet he retained a sense of awe about the very process of freezing and thawing a living embryo and then using it to grow a normal animal. "I'm still surprised that it works. I'm still bound to say, How can they survive that sort of treatment?"

Like Ian Wilmut, who said he worked with sheep only as a prelude to working with cows, Willadsen also did his embryological tricks first in sheep. The embryo-freezing work began in sheep and then he transferred it to cows.

When he succeeded in freezing, then thawing, a calf embryo and showing that it developed into a normal calf, Willadsen went to a meeting in Wels, Austria, to present his results. "One of the speakers stood up and said, 'It's a freak thing. They've only had one calf and that's all they're going to get.' " But, Willadsen said, "he was wrong. There are no freak accidents."

Willadsen continued to focus on sheep whenever he wanted to try something new. "Sheep were sort of mini-cows," he said. They had more offspring than cows and were less expensive to care for. Their gestation period was shorter. And "one man could essentially do everything," in research with a herd of sheep, Willadsen said. At one point, he was nearly single-handedly taking care of four hundred sheep. He would bring orphaned lambs home with him at night, bottle-feeding them. And he devoted all of his waking hours to his experiments.

In the mornings, Willadsen would operate on the sheep, surgically removing fertilized eggs from their oviducts. He would work in the lab in the afternoons. In the evenings, he would give the ewes hormone injections and check to see which ones were in estrus, receptive to intercourse, which meant that their eggs were ripe and ready to fertilize. To do this, he used trained rams, who were taught to sniff the ewes and prepare to mount those that were in estrus but not to actually mount them. The rams, Willadsen explained, had learned that if they performed this duty they would be rewarded at the end of the evening by being allowed to actually mount a ewe.

In the summers, Willadsen tried wild and imaginative experiments. "We had lots of ideas and lots of energy. We didn't just work on sheep—we had all species of farm animals there. The things we published were far from all of the things that were going on. We tried everything, and many of the things we tried came to nothing. But you had the mental file."

He decided to try one of Hans Spemann's famous experiments from the turn of the century. Spemann had taken a hair from his newborn baby boy's head and tied it into a noose that he used to sever a fertilized egg, making identical twins. But Spemann had worked with the comparatively gigantic embryos of salamanders. Willadsen was working with the embryos of sheep, barely visible to the naked eye as white flecks. Working under a microscope, he used a silk thread and knotted it around the embryo. But he was stymied because he could not allow himself to crack the egg's shell, a springy, gelatinous coating called the zona pellucida that surrounds the egg. It was crucial to keep the zona intact because the only way he could grow the embryos to a stage where they could be implanted in surrogate mothers was to insert them into the oviducts of rabbits. The embryos grew well in the rabbit oviducts—unless their zonas were not intact. When that happened, the rabbits' immune systems would attack the embryos and destroy them. But that meant that Willadsen could never completely sever an embryo, and that bothered him.

One day, he hit on a solution. He would encase the embryos in agar, a jellylike substance made from seaweed, before he put them into the rabbit oviducts. No immune system cells could penetrate the agar, but the nutritive fluids from the oviduct could diffuse through it. He tried the agar method. "I started to get something that I was sure was going to survive. I was like a hunting dog seeing what he was sure was a pheasant. I was on a very hot trail."

Willadsen knew then that he could use his agar method to split a fertilized sheep or cow egg, grow the twin embryos in a rabbit until they were ready to implant in a surrogate mother, then freeze the embryos. Depending on when he thawed them, he could produce identical twins born at different times. This was, he said, a form of cloning since he was producing genetically identical animals. It did not involve the transfer of nuclei, and so it could never be used to clone an adult or even an advanced embryo. But it was, nonetheless, the creation of a clone.

This time, Willadsen went to a meeting in Germany to present his results. The audience's attitude, he said, was the same as that of his other audience, in Austria, when he had told them of his frozen calf embryos. They dismissed his success as a freak accident. Willadsen knew they were wrong.

The idea of embedding embryos in agar changed the picture for manipulating embryos of farm animals. "It meant that all the manipulations that until then had been very unsuccessful suddenly could be done," Willadsen told me. "You can't think of an embryo that won't grow in that system—all the livestock embryos and probably human embryos as well."

Willadsen was quick to seize on the new possibilities. He asked himself how many times he could subdivide an embryo and still get each piece to develop into a normal animal. If he had a four-cell embryo, could he divide it into fourths? Could he divide an eight-cell embryo into eighths? "I analyzed it in sheep, cow, goat, horse, and pig, with declining interest," Willadsen said. Subdividing embryos, though, could be useful for animal breeders. If you could take a valuable embryo and split it, you could multiply its worth. It was the same notion that later inspired embryo cloning, and it had the same motivation.

Animal breeders faced a quandary in their attempts to create animals with particular traits. The traditional approach is to inbreed animals that have been selected for certain traits and then mate those inbred animals, which tend to be feeble or runts, with healthy animals from the general population, thereby merging genetic selection with hybrid vigor.

But when the hybrids are mated, the valuable traits from the inbred animals tend to be diluted or lost. Embryo subdividing offered a solution: It could allow breeders to multiply the valuable animals, including hybrids, without the gamble of the genetic lottery. They could simply subdivide selected embryos, making multiple copies of a single hybrid creature.

There were, however, diminishing returns when embryos were subdivided. With sheep, for example, half embryos were just as healthy as whole ones—60 to 80 percent of those that were transferred to surrogate mothers resulted in the birth of lambs. Half of the embryos that were divided in fourths grew into lambs. But only 5 to 10 percent of the embryos that were divided into eighths survived. The story was pretty much the same in the other animal species, although, Willadsen said, "the pig was probably the least successful." On the other hand, "I wasn't very interested in pigs."

"From a scientific point of view, it was beautiful. From a practical point of view, the twin was the one to go for," Willadsen said.

Scientifically, the embryo subdivisions showed that even when an embryo consists of eight cells, at least some of its individual cells have retained the potential to become a complete independent embryo. They also showed researchers that they could now start to really manipulate embryos. They could take them apart and put them back together again without fearing that by rupturing the zona pellucida they would destroy the embryos' viability.

*   *   *

Willadsen, working with Carole B. Fehilly, who was later to become his wife, began making chimeras—animals that were made
by mixing together cells from two different embryos. He developed a special method for doing this. "I would grab the egg with
a pipette and with a small needle cut a ridge all the way around
the egg," slicing through the gelatinous shell that surrounds the
egg. Then he would pull the shell open, exposing the embryo
inside. He would remove single embryo cells by sucking them
out with a pipette so fine that only one cell could enter it at a
time. He would replace those cells with cells he'd removed from
another embryo, close the egg shell again, and allow the chimeric
embryo to grow.

Willadsen and Fehilly mixed cells from different sheep embryos
and they mixed cells from cow embryos. Then they took the next
step, mixing embryo cells from different species. In one experiment, they mixed embryo cells from a sheep and from a goat and
made fourteen sheep-goats. Willadsen also made sheep-cows, again
by mixing embryo cells. The animals that emerged looked like
sheep with spots, like those on cows.

Willadsen said he was not just making sheep-goats and sheep-
cows out of idle curiosity. His aim was to see whether he could
break the species barriers in pregnancy in order to breed animals
of endangered species.

His idea was to make multiple embryos by treating females of
an endangered species with hormones so they produced large numbers of eggs and then fertilizing the eggs in the laboratory—
relatively easy tasks. But because the endangered animals were so
rare and precious, there would not be enough females to serve as
surrogate mothers for the embryos. So, Willadsen thought, perhaps
he could use a female of a related species that was cheap and
plentiful as a surrogate mother. An animal like a sheep, in the
case of endangered wild sheep. An animal like a domestic cow in
the case of wild cattle.

That amazing notion led Willadsen to ask, What would it take for an animal of one species to carry a fetus of another species to term? If he made an embryo whose placenta cells were identical to those of its surrogate mother but whose fetal cells were from another species entirely, could that fetus be brought to term? For example, if the surrogate mother was a sheep and the placenta of the fetus was made up of sheep cells but the fetus itself was a goat, could it survive? Willadsen found that it could, thus paving the way for his idea of using chimeric embryos to breed endangered species. No one, however, has tried this method on endangered species.

Willadsen, in fact, said he was more interested in testing his fanciful theory by creating these chimeras than in publishing his data. But some German visitors to his lab found out about his work and went home and produced sheep-goat chimeras of their own. They sent their paper to *Nature,* which passed it on to Willadsen to review. "I said it was just plagiarism," Willadsen said and, in anger, he sent his own paper to *Nature,* which published the papers from both groups in the same issue but printed a photograph of Willadsen's sheep-goat on its cover. The creature was strange indeed. It had goatlike horns that were twisted like the horns of a sheep. Its coat was mostly the thick wool of a sheep, but it also had bands and patches of goatlike hair.

The embryo manipulation work led Willadsen naturally to the idea of cloning by moving nuclei of embryo cells into eggs. He could easily make twin embryos by splitting them, but, he realized, the only way he was going to efficiently multiply an embryo would be if he could take each cell of, say, an eight-cell or a sixteen-cell embryo and make from it a new embryo. That would require moving the nucleus of each cell of the embryo into an egg and letting the eggs develop into embryos, then fetuses.

A little later, Willadsen went to hear Karl Illmensee speak at Cambridge University about his claims to have cloned three mice by exactly the sort of nuclear transfer that Willadsen thought would be his next step. So, Willadsen said, "I tried to do it his way," but he was unsuccessful. "I don't know that it was Ill-mensee's fault. My techniques were not good enough." He used sheep and cow eggs—he wasn't very interested in mice—but the eggs were so small that they were traumatized by the blunt injec-tion of a new nucleus. The embryos died. Others in his lab who *did* want to study mice tried the experiment with mouse eggs. They, too, were unsuccessful, Willadsen told me.

Soon after that, Willadsen got into a public spat with Illmensee. It was at a meeting in France, where Willadsen was filling in as chairman of a session for a senior scientist at his lab, Christopher Polge, the man who had originally sponsored Willadsen. Illmensee asked Willadsen why he would want to clone farm animals when he could make twins so well simply by dividing embryos? Willad-sen tried to reply, but Illmensee seemed irritated by his answers.

"I said, 'Well, the professor's angry. Let's not bother him,' " Willadsen said. "Then he got up again and said, 'This is dumb. Why clone cows?' " Willadsen replied, "The professor is sulking," irritating Illmensee to no end.

Later, when Illmensee's results were being questioned, the Uni-versity of Geneva turned to Willadsen, thinking Willadsen's bit-ing comments meant he was suspicious of Illmensee, not realizing that arrogance and sarcasm can be Willadsen's way, and asked him to review Illmensee's work.

Willadsen thought it over. He was skeptical, he said, but he also thought that the only way to resolve the controversy would be for Illmensee to repeat the experiment. In science, the accused is presumed guilty until proven innocent. "In other words, you must be prepared to prove that you are right. You can choose not to, but then you can't blame them for assuming you are wrong," he explained. And so Willadsen told the commission that if he were Illmensee, he would simply repeat the studies, "and then the whole problem would go away."

After all, Willadsen explained, Illmensee and Hoppe were not saying that once, by an incredible, once-in-an-eon fluke, they had produced a mouse by transferring a nucleus of an embryo cell into an enucleated egg. Instead, "they were claiming that this was a basic rule, that mouse embryos could be cloned in that way." If they were right, they should be able to do it again.

"Let's assume that Illmensee was a swindler," Willadsen said. Then the question was, Why did he fool so many for so long? The reason, he thought, was quite simple: Scientists wanted to believe that it could be done. It made sense biologically. People felt it might well be true.

But, Willadsen added, "if you do a good nuclear transplant experiment and prove that cloning can't be done—ah, that's almost as good as doing it from a scientific point of view. It would have profound effects. You could almost argue that if it could not be done, then more major discoveries would be made than if it could be done." After all, if it was impossible to clone, even from the undifferentiated cells of an early embryo, that would imply that DNA is changed, irreversibly, within hours after an egg is fertilized. Willadsen said he never believed that. No matter whether Illmensee's claims to have cloned were right or wrong, Willadsen was sure that it could be done.

He decided to start with his favorite test animals—sheep—and to try to fuse a cell from an eight-cell sheep embryo with an enucleated sheep egg using an inactivated Sendai virus. The virus melts into membranes of cells. If you put two cells together with the virus in between, it will meld the membranes of the two cells.

Why eight cells? From his previous work subdividing embryos, Willadsen knew he was very unlikely to get a lamb using a cell from an eight-cell embryo. At the eight-cell stage, the embryo cells have already begun to specialize, even if they look identical. If this cloning worked, it would mean that the embryo cell's DNA, amazingly, had been reprogrammed by the egg: The clock of development would have been reset to that of a newly fertil-ized egg.

The Sendai virus fusion method did work—the cells fused and did not immediately die. But Willadsen did not like the method. "It was too messy," he said. "It was not a clean, beautiful proce-dure." He first had to grow the viruses in fertilized chicken eggs, which were easily contaminated with bacteria or other viruses. He then had to inactivate the viruses by exposing them to ultraviolet light, and there was the risk that this would not always be done correctly and live viruses might escape and destroy his enucleated sheep eggs. In addition, he said, it was not really acceptable to think of using viruses in experiments with livestock or, someday, with human embryos. But he stuck with it, even hiring an experi-enced technician from a laboratory at Oxford University to come and make what seemed like a lifetime supply of the virus.

Next, Willadsen tried another innovation. When it was time to do the actual merging of the enucleated egg with the sheep embryo cell, he used sheep eggs that had not been fertilized. Why? "Inspiration, I guess," he told me. "When I experimented I tried many things, but when I tried this, it worked immediately." Unfertilized eggs were so much more receptive to the nuclei he was adding that he has used them ever since.

Soon Willadsen was not only fusing sheep embryo cells with enucleated eggs but he was growing the embryos until they were ready to implant in the uteruses of surrogate mothers and he was implanting them, creating pregnancies. He knew others were working on the same sort of project.

"The jungle drums were going, people were trying to do nuclear transplants," Willadsen said. "Neal First was using fertilized eggs, but he was not making any progress as far as I could see. However, sooner or later, the news that I was having success with unfertil-ized eggs would leak out. So there was no time to lose."

Willadsen's two first lambs, cloned by transferring nuclei from eight-cell embryos into unfertilized eggs, were born in 1984. "The reality is that the very first experiment I did, which involved only three eggs, was successful," Willadsen said. "It gave me two lambs. They were dead on arrival, but the next one we got was alive."

Then Willadsen saw an ad in *Nature* for a machine that fused cells with a brief electric current. It was not designed with embryos in mind, but Willadsen tried it and discovered that it did the job. He discarded his Sendai viruses and never looked back. But sheer coincidence, Neal First in Wisconsin discovered the same machine.

In June 1985, Willadsen left the British lab to go to Texas, where a large land and cattle company, Grenada Genetics, wanted him to help establish a business cloning cattle. For his farewell party with his closest scientist friends, he roasted one of his sheep-cow chimeras. "It was not very tasty," he confessed, "but that was because the sheep part of it was too old—it should have been eaten when it was a lamb."

As he left, he envisioned himself embarking on a new adventure with a fledgling business but still, at heart, the person who dares to challenge scientific dogma. He would continue, he said, "checking fences," looking for holes in the scientific fabric, ways to break through boundaries that others considered inviolable.

8

THE ROAD TO DOLLY

Cloning mammals from adult cells will be considerably harder, but can no longer be considered impossible; it might be a good idea to start thinking how we are going to make use of such an option.

—DAVOR SOLTER

It was a time of great expectations. Suddenly, almost miraculously, in the waning years of the 1980s, a few scientists could clone whole animals from embryo cells. And companies saw gold. The idea was simple—it was exactly what W. R. Grace and Company executives had dreamed of when they financed Neal First's cloning work at the University of Wisconsin. Scientists would take precious cattle embryos, worth $500 to $1,500 apiece, divide them into their constituent sixteen or so cells, and slide the nucleus from each of those cells into an enucelated egg. The result would be sixteen embryos, each of which was a clone of the original one. They might do even better—cloning each of the sixteen embryos, making sixteen embryos from each of the sixteen, and so on, expanding one precious embryo into hundreds of identical clones.

And so the cattle-cloning business began. W. R. Grace and Company, which owned American Breeder Service in DeForest, Wisconsin, a leading breeder company for dairy cattle, was one of the first to take the plunge. The rolling hills of rural Wisconsin might soon be dotted with cloned dairy cows, its executives thought. Grenada Genetics, which had lured Steen Willadsen to Houston to clone for them, followed shortly, and hired some of First's students to join the nascent cloning team. The vast plains of Texas, where cowboys rode herd, could be the home of cloned beef. Genmark, in Salt Lake City, joined the burgeoning field, hoping to grow its cloned beef on ranches nearby.

It certainly was feasible to clone cattle from embryos. Steen Willadsen and Neal First had done it in Cambridge, England, and Madison, Wisconsin, respectively, and they had shown that the method was reliable and repeatable. In fact, using the same methods, researchers were soon able to clone horses, pigs, rabbits, and goats from early embryo cells. In 1997, shortly after Ian Wilmut announced the birth of Dolly, scientists in Oregon even cloned rhesus monkeys from early embryo cells, using the same methods, indicating that these close relatives to humans seem not much different from sheep or cows when it comes to embryo cloning.

As evidence of what lay ahead, Grenada Corporation released a photograph of three black Brangus triplets, identical bulls that were clones. The photo graced page 1 of *The New York Times*.

Willadsen dutifully cloned early-stage cattle embryos for Grenada, but at the same time, he pushed the boundaries of cloning once again. He decided that it was one thing to clone just to multiply valuable embryos. But the true test of whether enucleated eggs could reset the clock of a cell's DNA would be to use more advanced cells—those that were clearly, *visibly* differentiated. Although most scientists, like Neal First, thought that once a cell starts to specialize, its DNA cannot be brought back to its primordial state, Willadsen was not convinced. As far as he was concerned, the main obstacle to cloning older embryos was that it was so hard to tug the cells of those embryos apart. The older an embryo is, Willadsen said, the more tightly its cells stick together. "When you try to separate the cells, you are more likely to lose a bunch of them. They stick together so tightly that they'd rather die than be separated."

But difficult did not mean impossible. To satisfy his curiosity about cloning older embryos, Willadsen took cells from cow embryos that were a week old, consisting of 60 to 120 cells. The cells were specialized, forming an inner layer that would become the embryo and an outer layer that would become the placenta. Yet, despite the conventional wisdom that said these cells could not be cloned, Willadsen's cloning experiments worked: Live calves were born. Willadsen never published his result—he'd done the work only to prove to himself that he was right about cloning, he told me—and most livestock cloners had no idea he had done it. They remained convinced that it was impossible to clone differentiated cells.

Willadsen lasted only a year at Grenada, leaving in the midst of a patent dispute with the company's owners and joining another company, Alta Genetics, in Alberta, Canada. The owner of Alta Genetics made Willadsen a partner in the new business and Willadsen saw a chance to do what he wanted while making some real money. As an incentive, the venture capitalists who were funding the company promised to contribute an extra $1 million if he could make one hundred cloned embryos in a year. "We did it," Willadsen said, even though cloning so many embryos was a stupendous job for Willadsen and his small team. It meant that he had no time to try the daring experiments that broke the laws of nature. He was a businessman now as well as a scientist, and business came first.

Within a few years, however, the cloning business collapsed, for a variety of reasons. Companies were finding that it cost more money to clone than they expected. And several of the companies had management problems, compounding their woes.

Frank Barnes, who had worked on the cattle-cloning project in Neal First's lab, joined Grenada and then Genmark, and saw the demise of cloning in both places. "It didn't sell. That was the bottom line," Barnes said. "The technology worked, but the cost of the technology was more than the American farmer wanted to pay."

Although Alta Genetics was sparing no expense in getting embryos from the most valuable cows in the world, Willadsen said, the company found itself unable to market the embryos he had cloned. Willadsen was working feverishly, but the clones were going to waste. "I was cloning purebred calves that went to slaughter," he said.

Perhaps the only surprise was how quickly the cloning business crashed. Barnes was working at Genmark on the fateful day in May 1993 when "the investors decided to spend their money someplace else." And no, he said, he wasn't really taken unaware when that day came. But, he remarked ruefully, "I was surprised that they couldn't see that this was more than a three-year proposition."

With the demise of Genmark, the disintegration of Grenada, the decision by Alta Genetics to get out of the cloning business, and the move by W. R. Grace and Company to scale back its cloning efforts, highly trained scientists like Barnes and Willadsen had nowhere to go. University labs did not have support from the government or from industry to pay for cloning research, so there were few options in academia for the cloning specialists. Several drifted into the burgeoning business of in vitro fertilization, where there were limitless supplies of patients, the doctors running the clinics were constantly on the lookout for scientists who knew how to handle eggs and embryos, the future seemed secure and the money steady.

Barnes reluctantly left the cloning business, remaining in Salt

Lake City, where he had moved to work for Genmark, and commuting to jobs at in vitro fertilization centers on the West Coast. He would have liked to clone animals, but, he told me, "I decided there are easier ways to make a living. Investor capital was where it was at, and when they got tired of cloning, the business went away."

Willadsen left Canada and now lives in Florida, where he, too, works at in vitro fertilization labs. Neal First, who had a tenured position at the University of Wisconsin, focused on other problems in animal reproductive physiology. So did Randall Prather, who ended up at the University of Missouri, studying embryo development in pigs.

By the early 1990s, cloning had retreated as an issue. Few scientists took it seriously as a research project. It had become notable mostly as the subject of famous controversies, like the still troubling Illmensee affair and the notorious dissembling of Rorvik. Mouse researchers had moved on to other projects and had learned to create genetically tailored mice without cloning. Ethicists no longer spoke of cloning as a threat to humankind.

Those who had dreamed of riches lowered their expectations and settled for a steady job. Only Willadsen ever made a living at cloning. As a partner in Alta Genetics, he received a large new house in Alberta for staying with the company for two years—he still owns it—and he had stock options that he cashed in when the company went public, selling his shares for seven dollars each. They are worth less than half of that now.

"I benefited very nicely," Willadsen said. "One could say I am probably the only person in the world who has lived virtually exclusively from income made from cloning for the majority of my scientific career," he proclaimed, then stopped himself, remarking, "It's kind of a brash thing to say, isn't it?"

Soon, the only reminder of cloning was in the movies. Cloning was a premise for a popular comedy, *Multiplicity,* in which a man who had too much to do had multiple (adult) copies of himself made. It was the theme of the cult science fiction movie *Blade Runner,* released in video in 1992, in which clones were created to be drones on other planets. (The experiment backfired, however, when the clones revolted and threatened their creators on earth.) The dinosaurs in the blockbuster science fiction movie *Jurassic Park* were cloned, although the word *clone* was not commonly used to describe them. But while some members of the public wondered if it might someday be possible to bring back the dinosaurs, as happened in *Jurassic Park,* scientists quickly disabused them of that notion. Cloning had returned to the realm of science fiction, not science fact.

Yet at the very time cloning seemed to have finally disappeared from the scientific landscape, and when it was retreating even from popular culture, Ian Wilmut and his colleague Keith Campbell in Scotland were quietly taking up the cloning problem. Few scientists knew anything about these men or what they were doing. They were not working at a major university or even at a high-profile company, but instead were ensconced at a little-known research institute whose low brown buildings were built in the midst of grassy fields where sheep graze. Townspeople who lived just a short walk away paid little heed to the Roslin Institute, except to complain that the animal odors from the pigs, sheep, and chickens were wafting into their homes on soft summer evenings. An asphalt sidewalk leads from the tiny town of Roslin past the Roslin Institute, but few stroll that way. A double-decker bus rattles past the institute every half hour on its way to and from Edinburgh, but few commuters disembark. For most Roslin residents, the institute is just some farm research place, with a red bar across its entrance and a guard dog that roams the premises at night.

Neither Wilmut nor Campbell had been part of the short-lived cloning boom of the late 1980s. Neither particularly wanted to multiply precious cattle embryos. But then again, neither man was an academic scientist, asking questions simply to slake a thirst to understand how an embryo develops. These two scientists who finally cloned an adult animal, defying the wisdom of many experts, were following a path of their own, supported by a company that wanted to make pharmaceuticals.

Ian Wilmut is, to all appearances, one of the most unlikely heroes of the cloning saga. Quiet, self-effacing, and cautious, he came to research by a circuitous route and to cloning because he was pretty much ordered to find a way to make animals with particular genetic traits.

Asked why he became a scientist, Wilmut traced his story back to when he was about ten years old and decided that he wanted to be a sailor. He was not a farm boy—he grew up in Hampton Lucey, near Warwick, England, the son of schoolteachers. But he met a sailor in his town and was captivated. For years, becoming a sailor was all he thought about, until he discovered at age fourteen that he could never fulfill that dream because he was color-blind. And so, Wilmut said, he began looking for careers that would allow him to spend his time out-of-doors. That led him to the idea of farming. To prepare himself, he started working on farms on weekends and holidays. He went to the University of Nottingham and got a degree in agriculture. But when he looked at his skills and talents realistically, with a cold eye, he doubted that he would ever make it in the farming business.

"I realized that I don't have a commercial knack at all," Wilmut told me. He decided to explore research careers instead. While he was studying at the University of Nottingham he had discovered another passion—embryos. He had come upon them in the summer before his final undergraduate year when he worked as a summer student at the British Agricultural Research Council's Unit on Reproductive Physiology and Biochemistry in Cambridge—the same place where Steen Willadsen was later to work.

Wilmut's adviser was the embryologist Christopher Polge, who would later become Steen Willadsen's adviser. In Polge's lab, Wilmut saw embryos for the first time. He was hooked, mesmerized by the tiny specks that were inevitably moving along a path that would turn them into living, breathing animals. The mysteries of embryo development, the sheer thrill of seeing life in the making at its very inception, and of manipulating these minute balls of cells that were going to turn into cows or sheep or pigs was irresistible to him. "That's the fascination," Wilmut said. Giving up on farming, he decided he wanted to go back to Polge's lab and get a doctoral degree, working with embryos if at all possible.

Wilmut loved that lab. Where Steen Willadsen saw a place where professors pretty much ignored their young charges, and where, by Willadsen's standards, many were unproductive and disengaged, Wilmut saw a place where the pace was just right and the professors inspiring. "It was just a fabulous, very relaxed place," he said. He ended up getting his degree for work on the freezing of boar sperm, a project selected because there was a commercial interest in it. It could allow farmers to store the sperm of prize pigs and use it later for artificial insemination. The intrusion of crass commercialism did not bother Wilmut. "I am very, very comfortable with it," he said, explaining that he enjoyed the lab's tradition "to try to understand the basic science but do something useful as well."

So Wilmut stayed on for a few years after he got his doctoral degree, studying ways to freeze cattle embryos and even getting one calf born from a frozen embryo (he named it Frosty). Like Willadsen who came after him, Wilmut was a Milk Board fellow. But he never thought of himself as a person who would break the laws of nature or who would creep around the fences of the possible in science looking for holes, as Willadsen did. Instead, Wilmut settled happily into a trial-and-error type of research, typified by his method of discovering how to freeze boar sperm. "We had a thing we called a four-seater loo" for testing different freezing regimens, he said. It had four wells, into which thermos flasks containing embryos could be lowered and cooled at different rates. That was how he found the best method to freeze the sperm.

In 1973 Wilmut's fellowship was almost up and he had to find a permanent position. He and his wife, Vivian, toyed with doing something really adventurous—going to Australia and starting a new life there. Wilmut found a job in Australia, but, he said, there was a six-month gap between the end of his job in Cambridge and the beginning of the new job. He decided not to go. Thinking back, a look of puzzled wonder on his face, he said, "We were very conservative. It never occurred to us to just use our money and sail slowly around the world."

Instead, he found a job at the Animal Breeding Research Station in Roslin, Scotland, which was later to become the Roslin Institute. In October 1973 he arrived, ready to do applied animal science. He relished the rural setting—the fields dotted in the springtime with dandelions and clover, the quiet town of Roslin just seven miles from the ancient city of Edinburgh. In the distance were the Scottish hills, where he would take long walks. And within the mazelike corridors of the research building were the labs where he could explore the mysteries of embryo development.

While Wilmut's wife spent her days taking care of their three children and looking after a play group of other people's children, Wilmut began his scientific career. His salary was always modest—when he created Dolly he was earning only $60,000 a year and stood to gain very little even if his methods were commercially successful. Even the cloning that resulted in Dolly would provide him with at most $25,000 a year. But he was content as he became known among animal scientists as a solid, honest, hardworking researcher. His publications, the coin of the realm in science, came out steadily in respectable journals. His reputation was secure.

Wilmut's first project was to investigate why sheep and cattle embryos often died before being carried to term. It was an important problem for the animal industry, which saw large losses when animals miscarried; many of those miscarriages occurred naturally, even when no embryos had been manipulated by human hands. It also was a problem for medicine in general, since large numbers of human embryos also are miscarried. Wilmut had plenty of ideas about why the embryos might die, but he soon realized that testing his hypotheses, and getting results that were meaningful, would require huge experiments, involving hundreds of animals. "The question was technically very, very difficult to answer," he said. "We talked ourselves into a practical corner." Nonetheless, he persisted, studying the problem and ways to attack it until 1981, when the Roslin Institute shut the project down. They had conducted a major reappraisal of the center's research programs and decided that Wilmut's project was not worth pursuing. Wilmut was told to work on a gene-transfer project that required injecting genes into embryo cells. "When I say I was told, I was told," Wilmut said. He complied.

The new project seemed a far cry from cloning, but its limitations and frustrating aspects eventually led Wilmut to conclude that cloning was the best way to get the results he wanted. The idea was to make genetically engineered animals by adding genes to newly fertilized eggs. In that way, a scientist like Wilmut might, for example, add a gene for a protein like insulin to a sheep's cells and arrange it so that the gene was only turned on in udder cells when the sheep made milk. The result would be a sheep that produced insulin in her milk.

Adding genes to fertilized eggs requires endless hours sitting hunched over a microscope, slowly pushing a syringe containing the genes into the microscopic embryo cells. The experiment would almost never succeed since only about one embryo in five even survived the injection and became a lamb—the rest died of the trauma or died of natural causes along the way. Only one cell in one hundred takes up the added gene, and even those that do usually do not maintain it and use it in all of their cells.

Since he has a slight tremor, Wilmut could not actually do the egg injections. He was just as glad. "It was the one good thing to come from my tremor," he told me. Instead, he worked with a newly hired scientist, helping him with the difficult task of finding the translucent nuclei inside of newly fertilized sheep egg cells.

It was tedious work. There had to be a better way, Wilmut reasoned. Perhaps, he thought, he could grow embryo cells in the lab, add genes to them, then merge those embryo cells with intact embryos, creating a mosaic embryo much the way Steen Willadsen created chimeras—his sheep-goats and sheep-cows—by mixing embryo cells.

The idea was to add genes to laboratory-grown embryo cells en masse, using vast sheets of cells and flooding the cells with genes instead of injecting fertilized eggs one by one, thereby increasing the chances that some genes might be accepted. Cells normally are impervious to genes floating around outside of them, but you can add the genes attached to calcium phosphate, tricking the cells into swallowing them as though they were salt granules, or you can shock the cells with an electric current, forcing them to briefly open the pores of their outer membranes and let in the genes, or you can hide the genes in a slick lipid capsule so that they slide into cells. Once you add genes to cells, you can use a biochemical test to pick out the cells that are using the genes— the test would be the equivalent of a sieve that would pick out only those cells that were wanted.

In principle, it sounded simple. Molecular biologists had already discovered all the tricks for working with cells that were growing in laboratory dishes. And once you had embryo cells that were using the added gene, your work was almost done. All that was left was to replace some of the cells from normal embryos with these genetically engineered embryo cells and let nature take its course.

But Wilmut's project was more difficult than it sounded. In order to add genes to embryo cells, he would have to grow millions of cells in dish after dish and keep them alive and keep them pristine—ready to merge with a newly formed embryo and take up residence. When scientists tried to grow early embryo cells in the lab, they found it was tricky. The cells died or they changed into something else, a cell more akin to a skin cell—long and stringy with tendrillike appendages—than a smooth, spherical embryo cell. Once the cells changed, they were useless. You could mix these cells with embryos, but the embryos would not accept them and they would never become part of the developing creature.

There was only one species in which such experiments with embryo cells worked—mice. When cloning had failed in mice, researchers discovered a trick with early embryo cells that was almost as good as cloning for their purposes. That discovery led the legions of well-funded mouse researchers to abandon all attempts to clone.

The mouse researchers had a very different motivation from that of scientists like Wilmut. They had no interest in making farm animals with added commercially useful genes.

The mouse specialists, in contrast, wanted to use genes to make mice with conditions that mimicked human diseases so that they could study treatments in these laboratory animals. For example, to see if a human gene really contributed to high blood pressure, they might add the gene to a strain of mice and see what happened to their blood pressure. They also wanted to see what the function of genes was by adding them to a mouse or deleting them from a mouse and seeing what happened.

If nature was fair or even consistent, the mouse specialists should have had the same problem that plagued Wilmut—they should have found it impossible to grow early embryo cells in the lab and add genes to the cells or delete genes from them. But the mouse experts had found an answer—the so-called ES cells, the abbreviation for a miraculous group of cells known as embryonic stem cells.

ES cells are simply cells from sixteen-cell mouse embryos that flourish in the lab and never change from embryo cells to skin cells. To keep ES cells in their original primitive state, scientists add a protein called LIF to the nutritious soup that they feed the cells. LIF originally was found in embryos but now it is available from mail-order catalogues.

Alternatively, researchers can grow ES cells on top of a layer of embryo cells that supply LIF or proteins like it. These embryo cells had been irradiated, breaking their DNA, rendering them unable to divide. For a period of time, however, they still keep up their biochemical reactions, including making LIF and, some scientists suspect, other as yet unidentified proteins that ES cells need.

Biologists can add genes to ES cells or delete genes from them and then merge the altered ES cells with another early mouse embryo. The ES cells will grow as if they had always been part of that embryo. The embryo will be a mixture of ES cells and cells from the original embryo, and the resulting animal will be a chimera—some of its cells will be derived from the ES cells and will carry the genetic alterations induced in those cells. Other cells of the animals will be derived from the unaltered embryo cells.

Mouse researchers could even create animals that were not chimeras but that were made, instead, of cells that were entirely derived from the genetically altered ES cell. All they had to do was make a mouse embryo, remove the ES cells, grow them, and add the extra gene to the ES cells. Then they would find the one in a million of those cells that took up the gene and was using it, and add that genetically altered ES cell to another mouse embryo to create a mosaic mouse. Some cells and tissues of the mosaic mouse would have the added gene and others would not. Then they would search for a mosaic mouse that had the added gene in its ovaries or testicles. All of the mouse's sperm or all of its eggs would have that gene and so would all of its progeny.

Once you find a male with the gene in its testicles or a female with the gene in her ovaries, you mate that mouse and, presto, you have offspring that have the gene in every cell of their bodies.

Yes, it was tedious. Yes, it was kind of clumsy. But it worked, and it had transformed mouse molecular genetics. So Wilmut and other scientists throughout the world began spending their time looking for ES cells in other animals. The search went on in dozens of labs, for years. No one succeeded. Maybe it was a technical problem—scientists may just have failed to find the LIF of sheep cells or cow cells, the magic growth factor in these embryos that would allow their cells to grow in the laboratory and not differentiate. Whatever the problem was, neither Wilmut nor anyone else could solve it.

There was, however, yet another approach. One that sounded much simpler, if only it would work. Perhaps Wilmut could take older cells, from fetuses or even adults, grow them in the lab and add genes to them. Fetal and adult cells are easy to grow in the lab and it is simple to get large numbers of them. The problem is that you can't add these older cells, which already are specialized, to an early embryo and expect them to simply become part of that embryo, as the ES cells do. The only way around that problem was to clone the older cells once their genes had been altered. The drawback was that most people thought that cloning from any but unspecialized early embryo cells was impossible.

As the long golden Scottish summer days melted into the short chilly days of winter, as year after year went by, the gene injection project was as tiresome as ever. The elusive ES cells of sheep, if they existed at all, remained elusive. There seemed to be no way around the utter tedium of injecting genes into embryos.

Finally, in 1986, Wilmut heard a rumor that changed his life—

he learned that Steen Willadsen had cloned from 60- to 120-day-old cattle embryos. Wilmut had flown to Ireland for a scientific meeting. When the day's talks were over, he went to a local pub, joining other scientists in idle banter and shop talk. He struck up a conversation in the bar with Geoff Mahon, a veterinarian who had worked with Willadsen at Grenada Genetics, who confided what Willadsen had done. Immediately, Wilmut realized what the results meant and reasoned that if Willadsen had cloned from older embryo cells, it might be possible to clone from even more advanced cells, from fetal cells or even from cells from an adult. In other words, there might not be any biological barriers to cloning. "I thought if that story was true—and remember, it was just a bar-time story—if it was true, we could get those cells from farm animals," Wilmut said.

"I came back all lit up," Wilmut told me. He flew back to Scotland with the Roslin Institute's research director, Roger Land. Sitting side by side in the cramped airplane seats as the plane passed over the Irish Sea, Wilmut and Land excitedly planned their next moves. "I told him, 'We've got to look into this.' "

But what if it were not true? After all, Willadsen had never reported that he had done that cloning experiment, and he was the sort of wild and crazy scientist who inspired apocryphal stories. Before he took off down Willadsen's path, Wilmut decided that he had to visit Willadsen in person and find out if what Mahon had said was true. So Wilmut went to Canada and tracked Willadsen down in his lab at Alta Genetics. "He was very helpful," Wilmut said. "He said it was true."

On October 10, 1987, Wilmut had his first meeting with commercial sponsors who, he hoped, would support the cloning research. Their first reaction was extreme caution, Wilmut recalled. This was a time when the cattle-cloning companies were still beating their chests and proclaiming that they would fill the world with cattle cloned from embryos. Scientists who had a long track record, like Steen Willadsen and Neal First, were cloning. "There was a feeling that it was a rapidly moving area. One question that was asked was, Was it worth our while even starting?" North American scientists were already so far ahead, the sponsors said. Wilmut argued that he knew of no one doing such work and that, of course, the experiment was worth doing. Of course he could pull it off, no matter that people like Willadsen had been working on cloning for years. "Scientists are arrogant, aren't they?" he asked wonderingly.

Although Wilmut convinced his sponsors, it took another few years to get the project going. The Roslin Institute was shrinking and the commitment that Wilmut got was not firm. During the time he waited, he "spent a lot of time filling in pieces of paper," he recalled, writing proposals and explaining his plans. And he continued the slow and tedious work of injecting embryos with genes. "I was still doing gene transfers until 1990," Wilmut said.

Finally, the money came, and Wilmut set out to hire a cell biologist with a particular expertise. A doctoral student at the Roslin Institute, Lawrence Smith, had tried a few cloning experiments before Wilmut had had his talk with Steen Willadsen. Smith had noticed that the success of cloning seemed to be related to what is called the cell cycle. Growing cells follow a certain pattern that involves proofreading their DNA for mistakes in its genetic code. Immediately after cells divide, they enter a stage called G1, for "gap 1," when they check to make sure that their DNA is intact and when the cells start to enlarge, adding to their bulk. Then the cells enter a new phase, called S, for synthesis, when they copy their DNA in preparation for dividing. The S phase is followed by a phase called G2, for "gap 2," when the cells check their DNA for mistakes that might have occurred when they copied the DNA. The cells also grow larger in the G2 phase. The final stage is M, for mitosis, when cells actually divide in two and the process starts again.

Smith had proposed that the success of cloning may depend on where cells are in the cell cycle. When scientists pluck a nucleus from a dividing cell and inject it into an egg whose nucleus has been removed, that egg's DNA is in the midst of its own cell cycle. If the cycle of the newly added DNA is out of synchrony with the cycle of the egg's DNA, the egg may be unable to accept its new DNA and will not be able to set its clock back to the beginning of embryo development. This could account for the high failure rate of cloning attempts.

The cell-cycle hypothesis was intriguing, Wilmut thought, but Smith, a veterinarian, was not equipped to investigate it further. "Lawrence thought the cell cycle was the key to getting cloning to work," Wilmut said. "He clearly established that it was important. But he didn't understand the cell biology of it." So as soon as Wilmut had some research funds, "one of the first things I did was to employ a cell-cycle biologist. He didn't have to be an embryologist. All I needed was someone who knew about cell cycles."

He found Keith Campbell, a wiry Englishman with wavy brown hair that brushes his shoulders, who describes himself as "a keen hillwalker and off-road cyclist" and whose entire scientific career, once he established himself as an embryologist, seemed to be preparing him for this very moment.

Keith Henry Stockman Campbell grew up in the industrial city of Birmingham and trained after high school to be a medical technician. He began to work in a hospital lab, testing sera, tissue, and blood for bacteria and viruses. "It was really interesting for the first four years and I learned a lot," he said. But it soon became routine, he added, explaining, "Once you've learned it, you weren't developing anything." Campbell hungered for something more intellectually challenging. The day he qualified for his license as a medical technician, he quit.

He went to the University of London, getting a bachelor's degree in microbiology. "Then I decided to have a rest from academia," Campbell told me. He went to Yemen, to work in a pathology lab. He liked the work while it was novel and he liked the exotic setting, but he soon became restless again, looking for something else to do.

Campbell returned to England in April 1979, where he worked for a year on a Dutch elm disease control program in Sussex, identifying diseased trees and writing contracts for people who were going to destroy the infected trees.

Soon, he decided he wanted to go back to school and get a doctoral degree. He did it part-time, working as a research assistant to pay his way. His research was on cell growth and on understanding the cell cycle. His animal was the frog.

As he worked on frog cells, Campbell was also thinking about cloning and what it would take to make it work. He was fascinated by John Gurdon's astonishing work in the 1960s on cloning in frogs. He also was inspired by Karl Illmensee. Campbell had heard Illmensee lecture in 1984 and was transfixed by his report that he had cloned mice.

In 1990, Campbell was a postdoctoral research fellow at the University of Dundee in Scotland, studying DNA replication in frog cells. He had always wanted to try to clone mammals, but it is very expensive to work with embryos from large animals and no one was paying for it. So he had been working with frogs, which at least were cheap. Among his experiments, he had added human DNA to extracts of frog egg cells to see whether the frog cells could copy the human genetic material. "I was amazed by the fact that the frogs wouldn't replicate the human DNA for a while," Campbell said. "What they actually did was to modify the structure of the human nucleus to resemble the structure of the embryonic frog nucleus." The frog eggs changed the coating of proteins covering most of the human DNA into a coating that would cover frog DNA, in effect making the DNA of humans look like the DNA of a frog embryo. Campbell thus became convinced that it was possible to trick an egg cell into using a foreign nucleus. If that was so, it should be possible to clone from fetal or even adult cells by tricking an enucleated egg cell into accepting a fetal or adult nucleus.

Campbell saw his chance that same year, when he spotted an advertisement for an expert on cell cycles to come to the Roslin Institute and work with Ian Wilmut. He applied immediately and was hired. Right away, he began studying the cloning problem. By 1993, he switched from working with cattle to working with sheep.

"Sheep in Scotland are very, very, very cheap," Campbell told me. "At times, a sheep costs approximately one percent of a calf. We can have one hundred sheep for the price of one cow."

Scientists were still dreaming of finding ES cells in cows and sheep, Campbell said. "But I knew John Gurdon's work, and I said, 'You don't need ES cells.' Based on that work with the frog, you should be able to get a wide range of nuclei from a wide range of different cells to work," for cloning. "It was just a matter of finding the right way to do it," he concluded.

Campbell set out single-mindedly to find the best time in the cell cycle to clone. First he had to know where cells were in the cell cycle. The easiest way to do that was to synchronize all of the embryo cells so that they were going through their cycles in lockstep. Then he could remove them at different stages in the cycle and try cloning.

Since it is difficult to grow isolated cells from early embryos, Campbell thought at first that he would try to use drugs to synchronize the cells while they were part of intact embryos. He failed. The embryos were balls of cells, and to be sure that the drugs he injected would penetrate through to the core, reaching every cell, he had to inject doses so high that they killed the delicate embryos.

The other option was to break the embryos apart and try to grow the individual cells in the laboratory, synchronizing them there. But that path had its own problems. Early embryo cells, as was well known by then, either die when they are isolated and put in petri dishes in the laboratory or they differentiate and turn into long, narrow cells sprouting twiglike appendages that resemble skin cells.

Campbell decided that he had only one option: to work on more mature cells. Cells from older embryos will grow well in the laboratory and so it was at least possible they could be synchronized. The question was, how?

One idea was to put all the cells into a resting state by starving them. Cells have a natural protective mechanism that they use when it would be folly to grow. They put themselves into a state of suspended animation, which biologists call G0, for "gap zero," to signify that gap in the cell cycle. Campbell reasoned that if he starved cells until they were on the verge of death, they would all be synchronized in the G0 stage. And they might be in the perfect state for cloning.

"When I thought about it, it was logical," Campbell told me. "The G0 stage of the cell cycle had been ignored for a long time, but it has been implicated in the process of differentiation." One hypothesis was that when cells differentiate, or develop into specialized cells, they must rearrange the proteins that mask most of their DNA. A specialized cell, like a liver cell or a brain cell, uses only about 10 percent of its genes. Those genes are the ones necessary for basic functions, like digesting nutrients, as well as those that are needed for the specialized functions that make the liver the liver or the brain the brain. The vast majority of a cell's genes are kept hidden beneath a cloak of proteins and are not used.

During development, a cell that is turning into a heart cell, for example, will need different genes than a cell that is turning into a kidney cell. Scientists suggested that during development, cells put their DNA into a sort of suspended animation while they reprogram the DNA. While DNA is actively being copied in preparation for cell division or while the copied strands of

DNA are being pulled apart from each other as the cell splits in two, there is no opportunity for the cell to carefully move proteins around, hiding some genes and revealing others. So, Campbell reasoned, perhaps the same sort of thing happens when the newly fertilized egg reprograms the DNA it got from the sperm to mesh with the egg's own DNA. Perhaps the egg slips into a resting state and rearranges proteins so that its newly combined DNA will be ready to orchestrate embryo development.

If that was so, Campbell decided, the best possible time to add the DNA of a cell to an egg, in an attempt to clone, would be when the DNA of that cell was resting. If the added DNA was in G0, it would already be prepared, so to speak, for the egg to reset its clock.

Campbell and Wilmut had enough money to work on the project for about two months. Since it was not the sheep-breeding season, it was impossible to get ripe sheep eggs for their study, so they began to work with cattle, which can be bred all year round. They starved fetal skin cells, forcing them to enter the G0 state, and then tried cloning. It seemed promising. These fetal skin cells were differentiated cells, so, in theory, they should not have worked at all in cloning. But the eggs took up the genetic material from the fetal cells and grew to the blastocyst stage. It was not a successful cloning experiment, since they never even implanted the embryos in a surrogate mother. But it was a first step.

Nonetheless, Wilmut said "we were still hankering for ES cells." The next October, they began again to look for those miraculous cells which worked so well in mice, making cloning unnecessary. Campbell took cells from nine-day-old sheep embryos and began growing them in the laboratory. At first they looked like ES cells, and he was encouraged. But then they changed, flattening and sending out branchlike appendages, turning into something akin to skin cells. They were, without a doubt, differentiated. And no one had ever before been able to clone from such cells.

Maybe, Wilmut reasoned, the thing to do was to define exactly the point when a cell growing in the lab could be cloned, and when it had differentiated to a state in which cloning was impossible. He decided to follow the same sort of methodical process that he had used when learning to freeze boar sperm with his "four-seater loo."

Wilmut analyzed the process of growing early embryo cells into three stages. First, the cells were intact in the embryo—they were smooth round cells that showed no outward signs of differentiation. After they had been removed from the embryo and grown in the lab for a short time, they looked like ES cells, still round and smooth but making proteins that were manufactured by differentiated cells and not by embryo cells. Finally, as the cells continued to grow in the lab, they differentiated, becoming long and stringy and resembling skin cells.

Wilmut knew it was easy to clone the first group of cells (those from early sheep embryos) and he knew it was supposed to be impossible to clone from the last group of cells (the differentiated skinlike cells). The question was, where would the cloning process break down? How long would it take, how differentiated would the cells have to become before cloning was impossible? "It would teach us how quickly things went wrong," Wilmut told me.

So they began. Campbell tried his trick of starving the embryo cells to put them into the G0 state before trying to clone them. Then he tried to clone, in turn, the cells that were taken directly from early embryos, embryo cells that had been grown for a short time in the laboratory and then starved to put them in G0, and starved embryo cells that had differentiated in the lab.

But something astonishing happened. To Campbell and Wilmut's complete surprise, it did not seem to matter whether the cells were taken directly from an embryo or whether they no longer even resembled embryo cells. As long as they had put the cells into the G0 stage, the sheep eggs accepted the cells' DNA and used it. The eggs turned into embryos.

Campbell was delighted, and he was beginning to suspect that he was about to make history. No one had ever before been able to clone from cells that had grown in the laboratory and changed their forms. Yet he had cloned three embryos from cells that had started to differentiate, flattening and looking like skin cells.

Campbell and Wilmut ended up with fourteen embryos that were ready for transfer to surrogate mothers. Six were cloned from early sheep embryo cells that had not been grown in the lab. One came from an embryo cell that was grown in the lab for only a short time but had not yet changed its shape—it still looked like an embryo cell. The remaining seven were from embryo cells that had flattened and differentiated in the lab and that looked like skin cells.

All the clones were from Welsh mountain sheep, a species that is white with curly fleece. The surrogate mothers were Blackface sheep. These animals, as their name implies, have black faces. Their fleece is long and shaggy and they are larger than the mountain sheep. If the experiment worked, the lambs that would be born would bear little resemblance to their birth mothers; they would resemble instead their Welsh mountain sheep genetic parent.

As the pregnancies continued, Campbell's excitement grew. Five of the fourteen pregnancies progressed. The fetuses had heartbeats, they moved in their mothers, the researchers could see them with ultrasound.

In the six weeks before the ewes were due to deliver their babies, Campbell began sleeping on the floor of his tiny cluttered office at the Roslin Institute each night, waking every hour to check the pregnant ewes. "We didn't want to lose those lambs," he said.

In July 1995, the ewes went into labor, giving birth to five lambs. Two died minutes after they were born. One died ten days later—it had a hole in its heart, a fairly common birth defect in Welsh mountain lambs. But the two that survived were clones of

9

TAKEN BY SURPRISE

It would have been naive to think it was possible to have
prevented this.

—IAN WILMUT

On a stifling Wednesday morning in June 1997, when the air
was thick and the temperature soaring, the American Association
for the Advancement of Science held its first-ever public forum,
an opportunity for private citizens to hear and question scientists and ethicists on a major issue of the day. The issue was
cloning.

Ian Wilmut was there in the association's sleek new building
on New York Avenue in Washington, D.C., looking jet-lagged
and with his head sinking to his chest as he sat on the stage in
front of a few hundred people. But when it came time for him
to speak, he was smooth and polished, and he obliquely raised
one of the central questions that has come to haunt scientists since
the cloning of Dolly: How could scientists not have known what
was coming? Why was this cloning such a complete surprise?

Wilmut asked the question indirectly, in the context of an
ethical issue. Over and over again, he told the group, ethicists,

politicians, the public, even scientists have said that they were horrified by cloning and that the research should have been stopped before it got so far. But, Wilmut asked, "when would you stop it?" Would you have stopped Robert Briggs and Thomas King from trying to clone frogs in the 1950s? Or would you have stopped John Gurdon from his frog experiments in the 1960s and 1970s? Would you have halted Steen Willadsen from breaking the laws of nature and cloning sheep and cows from early embryo cells? Or was the seminal event the births of Megan and Morag?

No, Wilmut concluded, "it would have been naive to think it was possible to have prevented this."

The problem is well known to historians; it is to define at what point the course of history is changed. It is a task that may be possible only in retrospect. To decide how important a world leader is, historians usually have to wait a few years, or even decades. To decide what event was epochal and what was not, they may have to observe how subsequent events played out.

One way to see this is to notice what news journalists play up, and what they ignore. For its one-hundredth anniversary in 1996, editors at *The New York Times* compiled a coffee-table book of the most important page 1s for one hundred years, an as-it-happened view of the major news events of a century. I found it endlessly fascinating. One of my favorite front pages reports the stock-market crash that ushered in the Great Depression. The page 1 headline on October 30, 1929, said STOCKS COLLAPSE IN 16,410,030-SHARE DAY, BUT RALLY AT CLOSE CHEERS BROKERS; BANKERS OPTIMISTIC, TO CONTINUE AID.

If economic, or even political upheavals, can be hard to appreciate while they are happening, science is infinitely harder to judge. Science can seem a world of its own, understood only by superspecialists. All too often, it seems, we are confronted by the spectacle of dueling scientists who themselves are often gambling on some half-understood concept. Salt raises blood pressure, some say. No, it doesn't, say others. The water for the earth's oceans came from tiny comets that are, in essence, celestial snowballs that rain down on the upper atmosphere, some say. Nonsense, others reply.

Only when the dust finally settles can we see the path that took us to the astonishing present. And as the story of the birth of Dolly illustrates, even people like Wilmut and Campbell, who were perhaps in the best position to know what was happening, somehow missed the enormous significance of what they had done.

Part of the problem in recognizing the path to Dolly was that so many scientists had convinced themselves that it was impossible to clone an adult. The work leading up to Dolly seemed to be complete unto itself, not a prelude to Dolly.

But the story is more nuanced than that. Another aspect is that most of the scientific community was looking the other way. As Steen Willadsen and Neal First at the University of Wisconsin, and the other animal scientists at companies and agricultural universities, showed that they could clone embryos, their work seemed to be so pragmatic. And these people were working in a backwater of science. As all scientists know, science is cliquish and trendy. Some fields of inquiry are hot, and every minor advance in those fields is trumpeted; other fields are all but ignored.

Even that seminal event—the cloning that led to the births of Megan and Morag—was not appreciated by most scientists or by many members of the press because it was outside of the mainstream and it took place not at an Ivy League university with an aggressive public-relations staff but at a tiny rural research institute where a guard dog named Buster roamed the grounds at night.

At *The New York Times,* for example, we were oblivious to the research. Not only was the birth of Megan and Morag not on page 1, it was not anywhere in the paper, not even in the weekly science section that appears on Tuesdays. The name Ian Wilmut had never appeared in the *Times*'s pages until February 24, when the birth of Dolly was announced.

Many scientists, too, failed to realize what was happening. Although Wilmut and Campbell had published their paper on the birth of Megan and Morag in *Nature,* and although publication in *Nature* gives a science paper about as much prominence as it is possible to get, many investigators failed to notice the paper or, if they saw it, to appreciate what had been done. Some, like Shirley Tilghman, the molecular biologist at Princeton University, who avidly reads *Nature* every week, skipped right over the paper by Wilmut and Campbell. Tilghman ruefully confessed that she never saw it because it dealt with sheep and few molecular biologists paid much attention to research involving farm animals.

Others, Campbell told me, "kept going on that these were ES cells." They insisted that what he and Wilmut had done when they produced Megan and Morag was to find the elusive ES cells of sheep and so, these scientists assumed, they had not really broken a barrier and cloned from differentiated cells. Wilmut had spent many long years searching for those cells, hoping that sheep embryos might have embryonic stem cells that were like those of mice. But ES cells are special because they remain developmentally uncommitted embryo cells even though they are grown in the lab. The cells Wilmut and he actually cloned from were fundamentally different—they were more like the cells of an adult.

Perhaps it was this atmosphere of mostly blind indifference to their work that gave Wilmut and Campbell the courage to try to clone an adult sheep. If the world had sat up and taken notice with the births of Megan and Morag, if the ethicists who started wringing their hands at the birth of Dolly could have stood up in public forums and asked, Should we take the next step?, it is possible that history would have taken a different path. After all, the real practical importance for Wilmut and Campbell was in cloning from fetal cells that could grow easily in the lab and that could be genetically altered. The cloning of an adult was more of a lark, a way to see if they were right that there were no longer any boundaries in cloning. But at the time, the cloning of an adult seemed to be a fascinating experiment, and so Wilmut and Campbell began writing proposals to get funds to actually do it.

"I think people thought I was mad," Campbell said. "I remember telling people at a meeting that my aim was to clone an adult and that I would do it within the next two years." Their reaction, he said, was "yeah, yeah."

Campbell told me, however, that he was supremely confident that he could succeed because he never thought there was any intrinsic biological reason that would preclude the cloning of adults. After all, he said, "we know that DNA is not lost. So I look at it like this: We've got all the bits for building something and we've got the instruction manual. But as the original vehicle was built, people translated the bits they didn't want into a different language. So it's a matter of getting all the information back together again, translating it back to the original language and putting the bits back together again." Of course, he confessed, "that's a very tall order," and no human being knows how to do it. And so, Campbell said, "I tried to think of ways of getting cells or eggs to do it for me without knowing what they were doing." Conventional thinking that would argue that such a feat would be impossible failed to persuade Campbell. "One advantage I had is that I don't believe what people tell me. I never did."

Wilmut got the funding to continue his cloning work, mostly from PPL Therapeutics, a struggling company on the grounds of the Roslin Institute that was formed in 1987 to commercialize the institute's products. The company was run by Ronald James, a gruff Scotsman who was trained as a biochemist and who knew what he was getting into with the cloning experiment, but who also saw an opportunity for the company to turn a profit.

James joined the firm in 1991, after working as a portfolio manager for Prudential Assurance's venture capital arm. He brought in Alan Colman as a research director. Colman was an embryologist who had worked with John Gurdon, the man who had cloned frogs from tadpole cells in the 1960s and whose work had inspired would-be cloners ever since.

When I visited James in the spring of 1997, the Roslin Institute still had the look of a fledgling company. It was housed in a small and new-looking gray building with teal-blue trim, and its tiny blue-carpeted foyer was piled high with boxes, apparently filled with recently delivered supplies. James's office was modest and had no secretaries guarding its door. Just outside his windows were fields where sheep grazed.

James told me he had discovered the Roslin Institute in the most roundabout way. It was 1989, and the institute was off to a slow start, with just six employees. Its research director worked only part-time. James, at the time, was a venture capitalist, looking for companies to invest in. On a flight with Colman to Geneva, James started talking about how there were riches to be made by any company that could figure out cheap, reliable ways to make valuable protein drugs, like drugs that could dissolve blood clots and that cost thousands of dollars per dose. "I don't know if it was the alcohol I'd been drinking or that I was high in an airplane," James said, but he thought of a wild idea— forcing toad eggs to make the proteins for him. He told Colman that since toads lay their eggs in long ribbons, it would be easy to simply take a string of toad eggs and zip down them with a microinjector, a pipette that could quickly zap genes into each egg. The genes would then coerce the cells into making the proteins.

Colman, James recalled, told him he was "stark raving mad." But, James added, "a week later, he called me back and said, 'Forget about frog eggs. What about chickens?'" Colman's idea was to inject genes into hen's eggs and induce the eggs to make protein drugs, like clotting factors used by hemophiliacs or even insulin used by diabetics, at the same time as they made albumin, a protein that is the major constituent of egg whites. To obtain the drugs, technicians would simply break open the eggs and extract the drugs from the egg whites.

It sounded just crazy enough to be brilliant, the two decided, and so Colman and James proceeded to spend the next eighteen months looking for scientists who would attempt the experiment with hen's eggs. He said they found a group "south of the border," in England, that seemed to have the expertise and offered them money but, James said, "they turned us down. They didn't think it was fair to take my money because they didn't think they could succeed. To this day, I can't understand it." Then James discovered that scientists at the Roslin Institute were working on chickens and were willing and able to try the experiment, so James and Colman decided to go there.

Soon, James said, he learned that Roslin scientists already were working on injecting sheep and cow embryos with genes to create animals that would make drugs in their milk. "You didn't have to be an intellectual genius to see that it was the same idea," James said. He managed to raise £10 million, or about $16 million, for the Roslin Institute from private investors and began working with a Danish drug company, Novo Nordisk, to produce animals whose milk would contain a protein that dissolves blood clots. PPL also merged with an American company, TransPharm, of Blacksburg, Virginia, that was trying to make cows that produced proteins in their milk.

By 1991, Wilmut had created one such line of sheep. They made the drug alpha-1 antitrypsin every time they made milk. The drug is used to treat an inherited lung disease. But James was persuaded by Wilmut's argument that cloning, if it could be made to work, would be a far better way of making animals that would be a sort of living drug factory. It would allow him to bypass the tedious, hit-or-miss method of injecting genes into embryos and hoping that, by sheer luck, an embryo or two out of thousands injected would take up the genes and use them to make drugs. When Megan and Morag were born, James and Colman immediately grasped the importance of the experiment, quickly filing for patents to protect their investment.

With James's support for the experiment to create clones of an adult sheep, Wilmut began looking for a collection of adult cells that grow easily in the lab. He came across vials of frozen udder cells that PPL had saved from a six-year-old sheep that had lived on another farm. "I have no idea what that original animal was being used for at the time," James said. No one seems to have followed the ewe's fate, and although it is pretty certain that she was killed years ago, no one knows exactly when or how she met her demise.

It sounds like a tale by O. Henry. What should be the second most famous sheep in history, the ewe that Dolly was cloned from, died in anonymity and probably was eaten by some unsuspecting Scottish family. Yet it shows, once again, how profoundly nearly everyone involved in the Dolly experiment misread its importance.

Only two minor details of the Dolly saga were forseeable. First, it was no accident that Dolly was cloned from a female. The sheep that graze on the Scottish hills are almost all females. Males usually are slaughtered for food when they are baby lambs; only a few are spared to be sperm donors for flocks of ewes.

And second, sheep experts will tell you that if Dolly was cloned from a six-year-old ewe whose udder cells were then frozen for three years, there is no way that that ewe was still alive at the time her clone, Dolly, was born. That aging ewe could not possibly have escaped her ignominious fate. Sheep, which have only bottom teeth, tend to wear them down over the long winters eating rock-hard turnips. By the time sheep reach the age of six, their lives are nearly at an end because they have lost their teeth and can no longer eat. It is uncertain, Wilmut said, how long a sheep would live if it was coddled and its teeth protected—there had been no reason even to ask that question—but the ewe that was cloned to produce Dolly would have been treated like any other sheep and "put down," Wilmut told me, around age six.

Not thinking that they might one day be plagued at every meeting with questions about what happened to that six-year-old ewe, Wilmut and Campbell planned their cloning experiment. The frozen udder cells were only a small part of it. It was going to be a complicated study, designed to wring the most information possible out of a single experiment. First, Wilmut and Campbell wanted to ask if it was possible to clone an adult, using fully differentiated adult cells. If such cloning was possible, as they suspected it was, they wanted to know how much harder it was than cloning a fetus or cloning an embryo. To make this comparison, they would try, in the same experiment, embryo and fetal-cell cloning. Finally, they wanted to assess two ways of growing the cloned embryos before implanting them in surrogate sheep mothers. Their preferred way to grow these embryos was in a sheep oviduct—the method that Steen Willadsen had used—but there was a law in Scotland that was passed, Campbell told me, under pressure from animal rights groups, that said that researchers could anesthetize an animal only once. That meant that they had to anesthetize a sheep and implant an embryo in her oviduct, then kill her when it came time to take it out to put it in the womb of a surrogate mother. The other way to grow those cloned early embryos was in a petri dish in the lab. It was much simpler and much easier to grow embryos this way, but scientists had the impression that that was not as successful as the sheep oviduct method. The cloning study would compare the two methods by growing most of the fetal and embryo cell clones in sheep oviducts and the others in petri dishes. The clones of adults would all be grown in sheep oviducts, the researchers decided, so as not to lose any of them.

For cloning, they would use three types of cells: cells from a nine-day-old sheep embryo, differentiated skin cells from a twenty-six-day-old sheep fetus, and adult cells from the frozen vials of udder cells. The embryo cells came from a Poll Dorset sheep. The fetal cells were from a Black Welsh Mountain sheep. The udder cells were from a Finn Dorset sheep, which had grayish-white fleece and a pure-white face. Using the method that had worked so well in the previous cloning experiment, Wilmut and Campbell grew the three types of cells in the laboratory before transferring their genetic material to eggs whose nuclei had been removed, starving the cells so that they would enter the G0 phase of the cell cycle and become quiescent.

It was laborious work. Wilmut and Campbell attempted to clone 385 embryo cells, and ended up with 126 new embryos. From 172 fetal cells, they obtained 47 embryos. And they got 29 embryos out of 277 udder cells. After growing the tiny embryos for a week in sheep oviducts or in the laboratory, Wilmut and Campbell transferred all of the clones of udder cells into surrogate mothers but only a fraction of the other cloned embryos, reasoning that it was not practical to use so many ewes in this experiment. In every case, the surrogate mothers were Scottish Blackface sheep—white sheep with black faces—so the mother would look very different from her baby if any of the cloned embryos developed into lambs.

In keeping with Wilmut and Campbell's suspicions, the embryos that were grown in laboratory dishes appeared not to fare as well as those grown in sheep oviducts. One out of five embryos cloned from embryo cells and grown in the laboratory was healthy enough to create a pregnancy in a surrogate mother sheep; fourteen out of twenty-seven grown in a sheep oviduct created pregnancies in surrogate mothers. The numbers were similar for the embryos cloned from fetal cells.

As soon as the embryos were transferred to the Scottish Blackface ewes, John Bracken, a specialist in monitoring sheep pregnancies, was put on alert. He began doing ultrasounds of the pregnant ewes two months after the animals had been in estrus. His first scans showed that 62 percent of the fetuses had been lost, apparently miscarried. But he also had good news for Wilmut and Campbell: One animal was still pregnant with a fetus cloned from an udder cell. Their attempt to clone an adult sheep was working, as long as this pregnancy held.

At 110 days after the pregnancies were begun, Bracken noticed that four more fetuses had died, all of them fetuses that were clones of embryos. In order to understand what went wrong with these fetuses, the scientists killed the ewes to extract the dead fetuses and examine them. Two fetuses looked fine and their deaths were unexplained; two others had abnormal livers, but no other birth defects. Meanwhile, the one fetus cloned from the udder cells remained alive.

As the day drew near for the sheep carrying that precious clone of the udder cell to give birth, Campbell and Wilmut grew nervous. They did not want to lose this lamb because of an obstetrical accident. But Campbell's wife insisted that she did not want Campbell to spend another six weeks of his life sleeping on the floor of his office as he had before Megan and Morag were born. She would not have him going in to the Roslin Institute at 11:30 each night and returning home at 6:30 each morning, after having awakened every hour to check the pregnant sheep to see if labor had begun. This time, Campbell hired someone to sleep in the lab.

In July, Campbell, confident by now that all was going well, went on vacation.

On Friday, July 5, 1996, the surrogate mother carrying the clone of the adult sheep went into labor. Bracken attended the birth, but did not call Wilmut. In part this was because the Blackface sheep that was the surrogate mother was of a species that was skittish around human beings. If a crowd gathered to watch her give birth, she might panic and the baby lamb might actually be harmed. But in part it was because, remarkable as it might sound in retrospect, Bracken did not appreciate what a momentous birth this was. And neither did Wilmut. Although Bracken called him as soon as the lamb was born, Wilmut's response was decidedly muted.

"It's bizarre but true that we didn't get lit up on the day Dolly was born," Wilmut said. "It absolutely sounds absurd now. I had actually bought some champagne," he added, but he never opened it. "One of the factors was that Keith was away and it seemed wrong to celebrate without him. This had been a team project." But, of course, it would not be the same to wait until after Campbell returned to celebrate. "Then you've lost the spontaneity," Wilmut explained.

Another reason for their strangely low key reaction, Wilmut told me, was that their success in creating Dolly had been so long and drawn out. First there was the idea of putting cells into the resting state known as G0, which Wilmut and Campbell thought was the most intellectually exciting part of the project. Then there was the creation of Megan and Morag, the proof in principle, as far as Wilmut and Campbell were concerned, that their idea would work. Then there was the transfer of the cloned embryo into a surrogate mother. Then there was the exhilaration of the first ultrasound that showed an apparently normal fetus with a strong heartbeat. Then, said Wilmut, "we were getting monthly, then fortnightly reports that things were okay."

Campbell said he was, of course, a bit anxious. He telephoned the lab every day when he was on his vacation to see if the sheep had given birth. But, he said, "by then I knew it was going to work. I always thought it would work once we had Megan and Morag. We were just proving a point I'd already made."

PPL wanted to patent the method that gave rise to Dolly, so Wilmut and Campbell agreed to keep the lamb's existence a secret until the application was filed. But as often happens in the gossipy science community where researchers travel the circuit of meetings, giving talks and hanging out together, rumors began to spread. In the fall, Alan Colman visited Brigid Hogan, a mouse embryologist at Vanderbilt University and an old friend, and told her what had happened, swearing her to secrecy.

Randall Prather, the scientist who had so fervently competed with Steen Willadsen to clone the first animal from early embryo cells, said he found out about Dolly independently at about the same time, from scientist friends. James Robl, another of Neal First's former students, also found out. He and a group of colleagues had dinner with Ian Wilmut a few months before Dolly's birth was made public. "We did our best to extract all his secrets," Robl said. "He didn't tell me, but while we were at the meeting, we used various strategies for getting secrets out. One of the best ways is to put the graduate students together."

Neal First found out from a postdoctoral student of his who attended a meeting in Paris in January and heard rumors of the lamb's birth. Another of First's former students, Mark Westheusin, who had gone from the short-lived cloning efforts at Grenada Corporation in Houston to a faculty position at Texas A&M University, said he'd heard about the cloning feat "through the grapevine. You go to meetings, people are sitting around the bar and having a few beers," and the word slips out.

But the rest of the scientific world, and the public, was shocked when Wilmut and Campbell's paper was published nearly eight months after the birth of Dolly. Lee Silver at Princeton was in the midst of writing a popular book on the future of biotechnology when the paper in *Nature* announcing Dolly's birth was published. He had written an entire section explaining that cloning from adults was biologically impossible. Galvanized and immensely excited, immediately grasping Dolly's implications, Silver stayed up all night rewriting his book. "Now there are no boundaries. Now all of science fiction is true," he told me.

Ronald James had anticipated a public frenzy when the *Nature* paper came out and had tried to prepare for it. "It was immediately obvious, certainly to most of us at PPL, that it would immediately grab attention and that the focus of that attention would be, Can you do this with humans? The thing that does it is that, superficially, it's a step toward immortality. And if you take a step toward immortality, everybody stops and takes notice." James said he was not surprised that the cloning barrier was broken; that's the way science is. "People once thought the world was flat. They once thought that mass was mass and could not turn into matter. It was thought that there was a force in organic chemistry, some vital force, that was needed to make an organic molecule," he said. All are now disproved and everyone has gotten used to the once shocking implications of those discoveries. The world would get used to cloning, too, he was certain, but in the meantime, it was best to prepare for a deluge of attention.

James hired a public-relations firm and started to plan his moves. He sent the paper to the British ethics committee, the Committee on Fertilization and Embryo Research, so that leading experts in the U.K. would be able to answer reporters' questions. He also sent it to the biotechnology trade organization in the United States and to a few U.S. ethicists. He planned a huge press conference for February 25, the day before the *Nature* embargo was up and journalists were officially free to write about the paper. In the meantime, the plan was for James, Wilmut, and Campbell to suddenly become unavailable so reporters would have to wait until the last minute to interview them. Then, on Saturday, February 22, *The Observer* published its story and all of James's carefully laid plans went awry.

At the Roslin Institute, however, just up a winding hill from James's office at PPL, the media frenzy took Wilmut and a few others by surprise. They knew what James was planning, of course, but they thought he was being unrealistic. Shaking his head at his own naïveté, Wilmut said ruefully that he thought the interest in Dolly might be somewhat less than the interest had been in Megan and Morag.

Patricia Ferrier, a good-natured Roslin Institute scientist who helped with the cloning experiments, said that no one had any idea that Dolly would be more than a passing sensation, at best. "When Dolly was born, initially we just didn't realize the full implications of how important it was." She and the others assumed that the journalists who began swarming through the Roslin Institute would soon go on to other things. "We thought it would be a week, and then it would be finished. Then it went on a second week, and we thought it would be finished, but no."

Of course, they knew Dolly was unique, Ferrier told me. Yet, she lamented, "we knew she was going to make an impact, but not that big an impact. We weren't ready for that amount of interest and we kind of got shocked by the possibilities."

Bracken said that on the day the news of Dolly broke, a friend of his was clipping sheep in Norway, at the top of a field. He heard about Dolly, Bracken said, shaking his head in amazement. "We weren't ready for that kind of interest," he said.

And the media onslaught hardly seemed to let up. When I spoke to Campbell at the end of May 1997, he said, "People are here every blinking day, it seems like."

Yet only Campbell among the scientists seemed to anticipate the way Dolly would be received. "Keith got it right," Wilmut told me, "but I don't know if even he quite saw the scale" of the reaction, he added.

When Dolly was born, she behaved like a perfectly ordinary sheep. But, Bracken told me, "as she became more and more famous, she rapidly became aware that she was different."

"We cosset her," Ferrier said with her strong Scottish accent. "All her needs are taken care of." Unlike other sheep at the Roslin Institute, who graze outside in the sunny spring days, Dolly is kept in a pen inside a locked cement-block building. She has never been outside, never eaten grass, but instead gobbles dark brown pellets of a sheep-food concentrate. Bracken explained that the scientists were worried about security if Dolly was let out, but they also wanted to protect her health. "Here she is in a controlled environment," he told me. "If we put her outside, she would be at risk of diseases spread from other animals and weather conditions, all of which could jeopardize a healthy animal." And no one wants to take a chance with Dolly.

For the first ten months of her life, Dolly shared a pen with Megan and Morag, but by May, she had to be separated from them because she was growing fat. Every reporter, every scientist who came by wanted to pet Dolly and to feed her, and she had learned to eagerly accept the constant plying of food. Megan, in the meantime, had gotten pregnant and had a baby, and Morag was pregnant, so they needed extra food while Dolly needed less. The only solution, the scientists decided, was to separate them from Dolly by a low fence.

Dolly knows she is special. Most sheep are wary of humans and retreat to the back of their pens when people come near. Not Dolly. She rushes to the front of her pen when visitors arrive, bleating loudly. When she was housed with other sheep, she would assert herself by turning over her trough as soon as she finished eating and planting her forefeet on it. There she would stand, chest puffed out, the queen of the pen.

The question now is, How much of personality is destiny? Is Dolly the clone of an egomaniac or is this an acquired behavior? Unfortunately, until another sheep is cloned from those same cells, which remain frozen at the Roslin Institute, we'll never know.

In the small town of Roslin just up the road from the Roslin Institute, the local citizens were baffled by the attention Dolly was getting. "A sheep is a sheep is a sheep," said Grahame A. Harris, proprietor of Ye Old Original Inn. Harris's modest stone inn on the corner of Main Street, with just six guest rooms, has never been a huge attraction—"I harbor no illusions," he told me. The media boom with the birth of Dolly did not make him rich. Day after day, local men sat in his inn's smoky barroom and chatted, and they expressed only bewilderment over Dolly.

But the rest of the world seemed to have understood all too well what Dolly means, and with a belated sense of history in the making, Wilmut does too. Now, however, he is finding it hard to think, to say nothing of plan new experiments, as he deals with an unending stream of requests—from politicians, from scientists, and from the public. Life is utterly changed for him and for Campbell. They have been peppered with letters and messages from people who want to weigh in on the ethics of cloning or, just as often, want a pet or a dead family member cloned. They are inundated with requests to give talks on their work to scientists. Wilmut was even summoned by the United States Senate to discuss the science and ethics of cloning. Both Wilmut and Campbell are trying to keep their hometowns and their home telephone numbers secret.

In order to concentrate on his work, Campbell takes the phone off the hook in his cluttered office. Wilmut has been dogged by reporters wherever he goes and has found himself making more transatlantic trips than he thought possible. He has tried to be gracious, and he remains awed by the people he has met or is about to meet. These include Senator Edward Kennedy, leading bioethicists like Arthur Caplan of the University of Pennsylvania, and a well-known gene-therapy researcher, W. French Anderson of the University of Southern California. He even had a run-in with Nobel laureate Walter Gilbert, who unceremoniously accosted Wilmut and demanded to know how Wilmut could be so sure that he had actually cloned a sheep from the cells of an adult. Laboratory accidents happen, Gilbert said, and it could be that Wilmut accidentally mixed up two batches of cells and that Dolly was actually a clone of an embryo. Wilmut, in his quiet way, was just as assertive and bluntly told Gilbert that there is no other sheep at the Roslin Institute that looks like Dolly; she is the only Finn Dorset and only those udder cells of all the cells in the laboratory were from a Finn Dorset sheep.

I sat with Wilmut on a bench in Central Park in New York

on a cool clear morning in the beginning of June 1997 and asked him about his new life, after Dolly. Wilmut, who had never been to New York before, was in the city for just one day. He had arrived the night before and was to leave that evening. He had come to talk to some investors who were interested in exploring the possibilities of using cloning to create cows that would be impervious to mad-cow disease. To help himself deal with jet lag, he was trying to remain on Scottish time; he had not even changed his watch. That morning, he told me, he awoke at 4:00 A.M. New York time (9:00 A.M. in Scotland) and started writing grant proposals.

Wilmut told me that he'd developed a plan to deal with the barrage of questions he was getting on the ethics of cloning adults and, in particular, adult human beings. He was trying to do what he called "marketing," by which he meant carefully choosing his words so as to avoid getting into the ethical debates about cloning humans. His fear was that because he is a self-described non-Christian, his arguments might be attacked because of who he was, for his beliefs, "or lack of them," he told me. He wanted to tread a thin line so that he did not encourage the cloning of humans but, at the same time, did not encourage an overly broad ban on cloning that might halt important scientific research. And so, he said, he told himself to keep a low profile and to refuse to be drawn into debates on what cloning might mean.

Wilmut said that he came unglued from the pressure only three times; on one occasion his wife helped him to settle down and on another occasion it was a public-relations specialist who helped him.

"I've got an image that works," Wilmut said. "Let's say you go walking on a gray, miserable day, you're going up a hill and you're tired and you think, 'What the hell do I do this for?' Then you go back home, have a shower, and say, 'This is really great.' It's like that." Every time he feels himself losing control and feeling desperate from all the pressures and all the demands on his time, he said, "I sit back and think, 'Well, I've met Ted Kennedy and I've admired that family at a distance for thirty or forty years.'" And he's met luminaries in the world of science and ethics that he would never have known if not for the creation of Dolly. So, he tells himself, "God help me, this is hard. But, on the other hand, what a fantastic opportunity."

1 0

THE PATH AHEAD

The potential benefits of cloning may be so immense that it would be a tragedy if ancient theological scruples should lead to a Luddite rejection of cloning.

—The International Academy of Humanists

Ian Wilmut would not like to see human beings cloned and he has said so repeatedly. He stood on a stage at a public forum on cloning and told the audience that when it comes to cloning a person, "I don't find it frightening. I find it sad."

Wilmut said he would worry about the pressures parents would put on a cloned child and he is offended by the idea of trying to create a copy of someone who has already been born. He gave me the same argument in Central Park when he spoke of how Dolly had changed his life. "Copying a person is what I don't like," he said.

Wilmut is not a professional ethicist or philosopher. His distaste for human cloning is clear, if not eloquently expressed. But similar views are voiced by many, including world leaders like President Bill Clinton, who stood in the Rose Garden outside the White House on a bright sunny Monday in June 1997 and said he found cloning "morally reprehensible." He later joined world leaders at a Group of Eight economic summit held in Denver, Colorado, and voted to oppose the cloning of humans, the first time an ethical question like that has come before an economic summit.

That widespread public distaste for cloning should give us pause, said Ezekiel K. Emanuel, a wiry young ethicist and cancer specialist at Harvard Medical School. Emanuel was a member of the presidential ethics commission charged with making recommendations on cloning. After ninety days of deliberation, the group asked for nearly the strongest possible constraint—federal laws to prohibit the cloning of humans with the possibility of reviewing the question in three to five years. Yes, Emanuel said, the arguments of cloning's opponents may not be well expressed. But, he said, the public outcry against cloning suggested to him that the problem is not with the arguments but with the inadequacy of our language to express people's visceral reactions. "The strong public reaction suggests a strong argument," he said.

"Is cloning inevitable?" Emanuel asked. "It's hard to say. But there is a difference between saying we should accept it or sanction it or turn a blind eye." Science, Emanuel said, does not have to be unfettered. Freedom of scientific inquiry is "an important American value," he said, "but there is no unlimited right to free scientific inquiry."

Yet as the idea that cloning could be possible sank in, a counter-reaction began to develop among some scientists and philosophers. Are we jumping to conclusions, they asked, letting our emotions sweep us in a direction that we might not want to go? What, in fact, are we afraid of and what benefits might we be giving up if we go too far with a cloning ban? Their views can sound almost heretical in the current atmosphere of fear and loathing of cloning. At least one scientist, Lee Silver, said he has found that many researchers are lying low, speaking only privately among trusted groups of friends. But Silver and a few other prominent academics have decided to take a stand.

In a sense, we are seeing a repeat of history, said Paul Berg, a Nobel laureate from Stanford University who was part of the epicenter of the recombinant DNA debate of the 1970s. When science accosts us with something new and frightening, like recombinant DNA or cloning, "the first reaction is fear." But then, Berg said, we should "reflect, review, look back at it," and see if our initial visceral reactions hold. And so he was "horrified" to hear Clinton declare his repugnance for cloning. Have we learned nothing from those difficult years when the nation grappled with the specter of recombinant DNA run amok?

The most star-studded group to take a pro-cloning position is the International Academy of Humanists. Its members, an assortment of eminent scientists and intellectual leaders, wrote an open letter calling cloning bans "the Luddite option" and urging that they be reconsidered. The signers included Francis Crick, a co-discoverer of the structure of DNA and a Nobel laureate, Richard Dawkins, an Oxford University professor of the public understanding of science, Herbert Hauptman, a Nobel laureate in chemistry, William V. Quine, philosophy professor at Harvard University, Simone Veil, a former president of the European Parliament, and Edward O. Wilson, the Harvard professor who is the father of sociobiology.

In its statement, published in the Summer 1997 issue of the magazine *Free Inquiry*, the humanists wrote: "The moral issues raised by cloning are neither larger nor more profound than the questions human beings have already faced in regards to such technologies as nuclear energy, recombinant DNA, and computer encryption. They are simply new."

Richard Dawkins contributed an accompanying essay posing the heretofore unthinkable. "But do you whisper to yourself a secret confession? Wouldn't you love to be cloned?" he wrote. He admitted that he would, out of "pure curiosity." He explained,

"I find it a personally riveting thought that I could watch a small copy of myself, 50 years younger." He would love to advise his younger self on mistakes to avoid in life. More seriously, he added, as he watched the cloning debate, he was struck by the fact that it has been more notable for its emotions than for its thoughtfulness.

The humanists' main concern was that an anti-cloning fever would result in Draconian laws that preclude the advancement of exciting science. "The potential benefits of cloning may be so immense that it would be a tragedy if ancient theological scruples should lead to a Luddite rejection of cloning," they said.

Those benefits of cloning, scientists say, may be what elevates the discovery to the realm of something that really does have the potential to change the world. If you ask scientists what they could do with the cloning discovery, they describe a plethora of possibilities that seem irresistible. Even Wilmut and others who say they never want to see a human cloned are eager to move forward and explore the new scientific landscape. Yes, they confess, the path beyond Dolly is still shrouded in fog, but they are able to make out a few outlines of what might lie ahead.

It is one thing to rattle off a laundry list of the potential benefits of cloning and then to counter them with possible harms to the human psyche. It is quite another to see the dreams of scientists described in full and loving detail by those who spend their days trying to break biological barriers. It is then that the science begins to shimmer with promise and it is then that the full impact of the cloning discovery emerges.

The most obvious but also what is likely to be the first economically important use of cloning would be to produce replicas of perfect farm animals and to use cloning to add genes to animals so that, for instance, a cow might make valuable drugs in her milk.

In fact, within five months of the announcement of Dolly's existence, and about a year after Dolly was born, Wilmut and Campbell took the first step along this path. On July 24, 1997, they announced the birth of Polly, whose name refers to the fact that she is a Poll Dorset lamb and a successor to Dolly. Wilmut and Campbell created Polly by growing skin cells from fetal sheep in the laboratory, flooding the cells with a human gene and a marker gene (a gene to help them identify the cells that have taken up the added gene), and selecting the fetal cells that best took up and used the human gene. Then they cloned Polly from one of those cells. Polly was born in early July, two other lambs with human genes were born toward the end of July, and the researchers also produced two lambs cloned from fetal skin cells that had just the marker gene, but not the human gene, added to them.

It was, of course, a vast improvement over the old way of making genetically engineered animals—the method Wilmut had slaved over that involved injecting embryos with genes, waiting for the embryos to develop and for the animals to be born, then looking for that one animal in one hundred that had the genes in its cells and that used them effectively.

"Instead of making the animal and then working out which animal is best, we wanted to do our selection at the cell level," Alan Colman, the scientific director of PPL, told me.

Cloning also allowed scientists, for the first time, to selectively remove genes as well as add them when they created genetically engineered animals. Removal of genes, Colman said, was crucial in producing animals whose organs might be used for human transplantation.

"In the case of the pig, it is quite well known that one major reason that pig organs are rejected by the human circulation is that there is a particular sugar on the surface of pig organs," Colman told me. Many investigators expect that if they could remove the pig gene responsible for adding that sugar molecule, then the pig organs would be more readily accepted by humans.

Lee Silver noted that the speed and the implications of the work are stunning. "After Dolly, everyone would have predicted this, but they were saying it would happen in five or ten years." The incredible thing is that it is happening so quickly.

"We are getting closer and closer to human beings now, too," Silver added. "All of this can be passed over to human beings. Genetic engineering of human beings is now really on the horizon."

And genetic engineering of humans, or even just cloning to make replicas of humans is, of course, at the heart of the cloning debate.

That is the issue that President Clinton asked his ethics panel to address. That is the application that Ian Wilmut so vehemently opposed in his every public statement. That is the reason religious leaders and ethicists felt their voices must be heard. And that, of course, is why the birth of Dolly was so shocking.

But scientists and ethicists who considered the possibilities of cloning soon found themselves confronted with nuances so subtle that many said they could not always make a hard and fast pronouncement. If we truly feel that it is unacceptable for scientists to experiment with human eggs, adding the genes of specialized cells to these eggs and seeing if they could get a clone to start to grow, then our society would have to give up research that might enable people to literally grow their own organs for transplant. That is an advance that would be hard to spurn. So maybe a sort of limited human cloning research is acceptable, some say. But then, where *do* we want to draw the line? Is it all right to do research that stops short of creating a baby that is a clone? Is it acceptable to do research that would end up with cloned humans in order to help couples who are infertile and desperate to have a baby of their own? Do we object if humans are cloned from early embryo cells rather than from the cells of an adult who is already born? How far do we want to go in policing in vitro fertilization centers? And if we cannot formulate clear, logical arguments against the cloning of humans, does that mean that attempts to prohibit cloning are a choice to pursue a Luddite option, as the humanists put it?

Even Ian Wilmut confessed that he could not draw firm lines when confronted with the nuances of human cloning research. And yet, if we really want to stop human cloning, it might be argued that any forays in this direction are tentative steps down a slippery slope.

To begin with the most stunning, but least disputed, possibility, consider the idea of using human cloning to grow your own organ for transplant. Scientists dream of starting with the body's largest organ, the bone marrow, which is also a liquid organ, so it would not be crucial for cells to arrange themselves in a specific pattern. Bone marrow makes the red blood cells that carry oxygen through the blood, the white blood cells that make up the immune system, and the platelets that cause blood to clot.

Stuart Orkin, a Harvard Medical School professor who studies blood cells and blood diseases, has never been one to hyperventilate over wild speculations. His reputation as an impeccable scientist whose work withstands the closest scrutiny has remained intact for decades. But, testifying before the presidential ethics commission on April 14, 1997, he predicted that it might be possible to use the cloning breakthrough to enable patients to grow their own bone marrow that would be a perfect match and ready when the patient needed it.

The idea would be to start the cloning process as though you were going to clone an embryo—take a patient's cell and fuse it with an egg cell whose own nucleus had been removed. The egg would reset the clock of the genetic material from the adult cell and what looked like a new embryo cell with the genes of the adult would start to divide. Now, Orkin said, comes the crucial step. You would add chemicals that direct cells to become bone marrow, thereby forcing all of these cloned cells, which have the potential to become any part of the body, to become bone marrow cells. The result would be marrow that was identical to the patient's own marrow.

Brigid Hogan, the mouse developmental biologist at Vanderbilt University, said she and others are edging in on being able to do this in mice—they have identified some of the biochemicals in mice that direct a cell's fate. But researchers are still a long way from being able to do this in humans. "There are species differences," Orkin said. "If one wants to get to an application down the line, it is important to do research with human cells.

"This is all very high tech," Orkin told the president's commission. "We don't know how to do it." But, he said, "the research we're talking about is the only way to learn whether it's possible. Most research is unknown—the only way to know if we can succeed is to begin."

Anyone who needs to know the importance of finding a way to make bone marrow has only to listen to leukemia patients, and their desolate families.

In June 1991, Jay Feinberg learned that he had acute myelogenous leukemia, a blood cancer. Feinberg, who lived in West Orange, New Jersey, was just twenty-two years old. The disease begins with a massive overproduction of mature white blood cells, which are the immune system cells that fight infections. The cells still function in this chronic phase of the disease, although there are too many of them, and patients are in no immediate danger of dying. Four or five years later, however, the disease enters a second phase that is deadly. Instead of churning out mature white blood cells, the marrow starts making enormous numbers of immature cells. The only treatment that works is to destroy the patient's bone marrow with chemotherapy while the person is still in the chronic phase of the disease and give the patient a new bone marrow that comes from a healthy person.

Feinberg knew that he had to find a bone marrow donor or he would die within years. The donor's marrow, however, had to be a very close genetic match to Feinberg. If it did not match, the new marrow would make white blood cells that would find Feinberg's own cells and tissues foreign and would attack them. The result, called graft versus host disease, is fatal—the patient is killed by an attack by his or her new immune system.

The first place to look for a donor is among brothers and sisters, then parents and relatives. But there are so many possible genetic patterns that some unfortunate patients discover that they have no relatives who can help them by donating marrow cells. That is what happened to Feinberg, and that is when the search for a marrow donor becomes wrenching, and expensive.

Feinberg began by searching an international registry of one and a half million people who had volunteered to be marrow donors for strangers. None matched. So Feinberg embarked on an international search of his own, testing more than forty thousand people at a cost of more than $2 million, most donated by strangers who heard of his plight and gave money to help him. Finally, years later, and just in time, Feinberg found a donor. But he was one of the lucky ones. Most who embark on a search, making their plight public, begging people to donate money or have their blood tested, spend months or years, and millions of dollars, for naught. Even worse, no marrow—unless it comes from an identical twin—is a perfect match, and so even in the best of circumstances, graft versus host disease is still a threat. A patient might finally find a donor, only to die as a result of the transplant. How fantastic it would be to jettison the risky transplants of other people's marrows and simply grow your own.

Even more fantastic would be to grow solid organs, like kidneys and livers. Ultimately, Orkin said, scientists might be able to bypass the initial step in which they add an adult cell's genes to an egg. If they could learn how the egg reprograms a cell's DNA, bringing it back to its primordial state, they might someday be able to force a cell to reprogram its *own* DNA and then differentiate into any sort of cell that the scientists want. That, of course, is the most futuristic scenario of all, Orkin warned, but it shows what might someday be possible. That process of learning to reprogram a cell's DNA would have to begin, however, with cloning.

Of course, the true fascination with cloning still revolves around the questions of why anyone would want to clone humans and what would happen if we tried.

Those, like Ezekiel Emanuel, who oppose human cloning, have said that no matter what you think about the morality of cloning a person, the fact remains that it is hideously dangerous. The Dolly experiment certainly gave grave hints of how ruinous cloning could be to most embryos, and no one could possibly expect to sacrifice hundreds of human eggs, embryos, even fetuses, to get one living clone. That alone, Emanuel and others on the president's ethics commission argued, would make human cloning unethical.

But Lee Silver likes to stun ethicists, theologians, and even many scientists by pointing out the errors of those who have argued from the Dolly experiment that cloning humans would be unsafe. He told me that he got up in front of scientists and ethicists at a meeting in Washington at the end of June 1997 and shocked his audience by arguing that cloning would be safer than the ordinary way of making babies by uniting sperm with eggs. The meeting was closed to journalists and to the public, in order to allow the scientists and ethicists to speak freely. Nonetheless, Silver told me, the pressure was on to say politically palatable things.

The first argument raised by those who fear cloning is that it would create monsters, genetic mistakes that are stillborn or, even worse, that survive to birth. How could we risk such a disaster? critics ask. But, Silver replies, cloning is actually genetically safer than normal sexual reproduction because it bypasses the most common form of birth defect—having the wrong number of chromosomes.

The vast majority of genetic birth defects occur because an embryo ends up with too many chromosomes, or too few. Jacques Cohen, who is scientific director of assisted reproduction at the Institute of Reproductive Medicine and Science at St. Barnabas Hospital in Livingston, New Jersey, told me how often these problems, known as aneuploidy, occur. "The rates are staggering," he said. As many as 40 to 50 percent of the eggs of women under age forty have one chromosome too many or one too few. "In older women, the rate is even higher, it may be nine out of ten eggs. That is why IVF [in vitro fertilization] is so inefficient and that's why older women don't become pregnant," Cohen told me.

These eggs with too many or too few chromosomes may be fertilized; similarly sperm with the wrong number of chromosomes may penetrate an egg and fertilize it. But the resulting embryos usually die almost immediately—the woman miscarries, often before she even knows she is pregnant. A very few such embryos survive to become fetuses and fewer still survive to birth, most often those fetuses with an extra copy of chromosome 21. These are the babies with Down syndrome. An extra copy of most other chromosomes is an invariably fatal condition, as is a missing copy of most chromosomes.

The chromosomal abnormalities occur when sperm and egg cells are produced. In the testes of a man and in the ovaries of a woman, progenitors of sperm and egg cells mature. As they do so, they divide repeatedly and develop into cells that have a single copy of each chromosome instead of the usual two copies. When that happens, some sperm and some eggs accidentally end up with one chromosome too many or one chromosome too few.

With cloning, such chromosome mixups cannot occur, Silver pointed out. After all, you are starting with a normal cell, from a normal adult, with the proper number of chromosomes. So the major cause of birth defects is ruled out.

The second major category of genetic birth defects, which is much less common than chromosomal abnormalities, is recessive genetic diseases like sickle-cell anemia and Tay-Sachs disease. They occur when each parent has a single copy of a gene that, in a double dose, causes a disease. The parents are healthy, but if their children inherit a mutated form of the gene from each parent, they will be ill. Cloning, once again, avoids this cause of birth defects because it starts with a cell from a healthy adult.

Some cloning critics have said that it is clear from the Dolly experiment that cloning is unsafe because Ian Wilmut started out with 277 eggs and ended up with a single sheep. But, Silver noted, only 13 of those eggs developed into embryos and 12 out of the 13 were miscarried early in pregnancy. That leaves a success rate of one in 13, already far higher than the success rate in the early days of in vitro fertilization.

The answer, of course, is that more research is needed to know how safe and how reliable human cloning might be. But the message, Silver emphasized, is that we certainly have no compelling information proving that it is dangerous and it is disingenuous to pretend otherwise.

The alleged physical danger of the procedure is just one argument against cloning humans, however. Another is that clones might look young but they would really be old—a newborn-baby clone would have DNA that was as old as the DNA of the adult whose cells were used to create the clone. Instead of having an expected lifespan of seventy or eighty years, the clone might live only as long as that adult had left to live.

This fear is related to a question that arose as soon as the existence of Dolly became known: How old is Dolly? Is she her chronological age, or is she really the age of the sheep from whose cells she was cloned? In other words, does DNA age and does the age of DNA determine an animal's—or a person's—life span? Or is the biological clock truly reset when a cell is cloned?

Those who say Dolly is genetically old cite an arresting image of DNA aging and cancer. At the ends of chromosomes are repeated sequences of DNA, like stutters. The idea was that these DNA sequences, called telomeres, were like ticker tapes that shortened every time a cell divided. They were supposed to be longest in embryos and to grow progressively shorter and shorter as a person aged. Finally, when the telomeres shrank down to nothing, the cell, and the person, died.

Cancer cells, however, were said to have telomeres as long as a newborn baby's, enabling them to divide forever and to be essentially immortal. The telomere hypothesis said that if scientists could lengthen the telomeres of the elderly, they might be able to restore youth. If they could shorten the telomeres of cancer cells, they might cure cancer.

If the telomere story was correct, then Dolly would not have long to live. After all, she was cloned from a six-year-old sheep, which is truly geriatric as far as sheep go. Scientist after scientist stood up after the Dolly announcement and cited the telomere hypothesis. Cloning could not be safe, they said; Dolly might look young but her telomeres were probably those of an elderly sheep.

Of course, there is one obvious flaw in this argument. More than 90 percent of all the cell divisions that ever occur in an animal's—or a person's—life occur in the womb, when a minute embryo grows into a full-term fetus. If Dolly's telomeres had only a few cell divisions to go, she could not possibly have made it through fetal life—her cells would have given out, their telomeres depleted.

Moreover, even if you disregard that theoretical argument, there is another problem with the telomere thesis: Eggs are packed with enzymes that lengthen telomeres. In fact, one of the first things eggs do when they are fertilized is to adjust the lengths of the telomeres on their chromosomes. So if a clone started out with telomeres that were too short, it is all but certain that the egg would lengthen them.

But even disregarding that flaw in the telomere argument, there are other problems with the thesis. All animals age, but different species have telomeres of very different lengths. Those with long telomeres do not live any longer than those born with telomeres that are shorter. Silver told me that mouse researchers have even created mice that do not have the enzyme that makes telomeres. The mice seemed healthy, so the scientists bred them with each other to see what would happen. The researchers are now up to their fourth generation of mice lacking the telomere enzyme, and they have yet to find anything wrong with the animals, Silver said. Their life spans are normal, he added.

Elizabeth H. Blackburn, a scientist at the University of California in San Francisco who is known as "the queen of telomeres" for her pathbreaking studies, told me of a discovery that is even more devastating to the hypothesis: Researchers have found that telomeres do not always shorten significantly with age and that cancer cells do not always have telomeres of a constant length. The telomere hypothesis, Blackburn said, is no more true than proposing that since everyone who grows old gets wrinkled, it is wrinkles that cause old age. "It was a black-and-white world a few years ago. But now the real world has intruded," she said.

As to whether Dolly was really six years old when she was born, no one knows. If she lives what appears to be a normal sheep life span of six or so years, it will be clear that her DNA's clock was truly reset. If she dies young, the answer to the aging question may not be clear. Perhaps, as Wilmut told me, we will just have to wait and find out.

Some cloning critics have said they worry about cancer in clones. Mutations in DNA accumulate over a cell's lifetime, and one leading theory of cancer says that a cell becomes cancerous when it garners mutations in several critical genes. If you clone from an adult cell, you may be giving that cloned baby the mutations of an adult cell. Instead of being most susceptible to cancer in old age, as most people are, the clone may be susceptible in childhood.

But there is a problem with that argument too, said Silver. The egg and sperm cells whose DNA normally combines to create the genetic material of an embryo also are adult cells. They are just as susceptible to cancer-causing mutations as any other cells in the body. So there is no reason to expect that a clone made from a skin cell or an udder cell, for example, would start out with DNA that was farther down the path toward cancer than a baby made the ordinary way.

All these arguments that cloning may not be so dangerous would be moot if no one ever intended to clone humans. But, at least in the private world of infertility clinics, a number of doctors said they were looking forward to using cloning methods to help couples have babies of their own—even, perhaps, babies who were actual clones.

Mark Sauer, an infertility expert at Columbia Presbyterian Medical Center in New York, said that what most intrigues him is cloning early embryo cells of humans, in much the way that scientists like Steen Willadsen and Neal First cloned the early embryo cells of cattle. He would like to take each cell from an early human embryo and clone it, making many identical embryos where there once was only one. He would implant some embryos in the woman's uterus immediately and freeze any extras for future attempts at pregnancy.

Yes, Dr. Sauer confesses, the woman might end up with identical twins, triplets, or even quadruplets, possibly born years apart. But the alternative might be no babies at all.

Dr. Sauer and others point out that if doctors could make many identical copies of an embryo, they could avoid having to give women powerful drugs, month after month, to whip their ovaries into overdrive. Doctors do this because they need as many eggs as possible, so that they can fertilize them and create as many embryos as possible, to increase the woman's chances of pregnancy. Since not every egg is successfully fertilized, and not every embryo develops beyond a few cell divisions (most, in fact, have lethal chromosomal abnormalities) and not every embryo that starts to grow in the lab successfully implants in the woman's uterus, doctors must maximize the odds by maximizing the woman's egg production. It would be much less expensive and much easier for a woman to have one or two embryos multiplied than to force her ovaries to produce a dozen or more eggs. The embryo cells could be cloned by adding their DNA to the DNA of eggs that would otherwise be discarded, because they failed to fertilize, for example. Or the eggs could come from young women who volunteer as egg donors.

It is such an attractive idea, said Robert Anderson, who is director of the Southern California Center for Reproductive Medicine in Newport Beach, that "I guarantee you that somebody, someplace, is working on it right now." In fact, he added, he guesses that doctors "in more than one place," are working on this sort of human embryo amplification.

But is it acceptable to those who say they are morally opposed to cloning humans? Ian Wilmut told me that he is not sure how to respond. "Multiplying embryos is a close one. It's very difficult to call," he said. He added that he would have no objections if the embryos were used all at the same time so that the woman might give birth to identical twins or triplets. He is not so certain he would approve of freezing some of the embryo clones and using them later. His fear is that a newborn identical twin of a baby that is already several years old would not be treated as a unique individual. On the other hand, Wilmut conceded, on the scale of moral objections to cloning, this one falls fairly low.

But that may be only the beginning, infertility specialists said. Jacques Cohen from St. Barnabas Hospital told me in May 1997 that he had already been contacted by three women, under age forty-five, whose ovaries have failed and who would like to have children. They wanted him to add the DNA from one of their husband's cells to an egg from a donor, whose own DNA had been removed. That would produce a clone of the husband. Alternatively, doctors might add the genes from one of the woman's cells to a donor egg, producing a clone of the wife.

For now, Cohen said, the method needs more work before it could be used in humans. "From a clinical point of view, you can only use a method like that if you feel that, based on animal work, the method is safe and efficient." And so, while he and others do further research, he has no immediate plans to clone anyone, although he has no ethical objection to using cloning to help people like the women who called him.

Alexander M. Capron, a lawyer and ethicist at the University of Southern California in Los Angeles, and a member of the National Bioethics Advisory Commission, said some infertility doctors have begged the commission not to prematurely ban human cloning. (In fact, without explicitly recognizing these doctors' pleas, the commission's final report suggested banning only the cloning of cells from people who had already been born, leaving the door open to many of the possibilities that the infertility doctors envisioned.)

One doctor called Capron saying he thought that, with more research on techniques, he could help women whose eggs could be fertilized but who always miscarried their fetuses. His idea was to create an embryo, then clone it by adding the nuclei of the embryo cells to enucleated eggs from donors who had had no trouble carrying a pregnancy and then implanting those embryos into the womb of the infertile woman.

"The doctor who told me about this was pleading. He told me that there are women for whom this is the only way to get pregnant," Capron said. The doctor insisted that the method should not be objectionable since the clones would not be made from adult cells. Therefore, he argued, this method was not in the same ethical ballpark as cloning from an adult cell, and he urged the bioethics commission not to ban it.

"I said, 'I think you're right. It doesn't raise the same problems. But in the minds of some, it raises the same issues. You will be making multiple copies of the same individual, not all of whom will be born at the same time or, possibly, even to the same mother,'" Capron said.

While the debate was going on, Jacques Cohen and Steen Willadsen were trying something very much like what this doctor had suggested to Alex Capron. The difference was that, while the doctor suggested moving the nucleus of an infertile woman's egg into the healthy cytoplasm of a donated egg, Willadsen and Cohen did the reverse: moving the healthy cytoplasm of the donor's egg into the egg of the infertile woman.

First they selected three women whose eggs could be fertilized normally but whose fertilized eggs failed to develop, indicating that there was likely to be something wrong with the eggs' cytoplasm. Each woman was matched with an egg donor, a young woman whose eggs fertilized and developed normally and efficiently. Cohen gave both the infertile women and the woman who was to be the egg donor hormones to synchronize their egg production. When both women produced eggs, an obstetrician removed the ripened eggs from the women's ovaries.

Cohen and Willadsen began by working on the donor egg, using a method that Willadsen developed. They put the egg in a solution that makes its membranes fluid, so they can fold in on themselves. Then they poked a large pipette into the egg, sucking out virtually all of the cytoplasm. The relaxed membranes folded around the bleb of cytoplasm, surrounding and protecting it. They were left with a ball of cytoplasm: an egg without its nucleus.

The next step was to merge this egg that lacked its nucleus with the egg of the infertile women, using a method much like Willadsen used when he was cloning cattle. He poked the ball of cytoplasm under the jellylike shell of the infertile woman's egg. Then he jolted the egg with a short burst of electricity so that the ball of cytoplasm would merge with the infertile woman's egg. Finally, Cohen and Willadsen injected sperm into the reconstituted eggs to fertilize them. (They could not simply let the sperm swim up to the egg and fertilize it because the electric jolt made the egg think it was fertilized—it no longer allowed sperm to penetrate it.)

Cohen told me that he and Willadsen tried this method with twenty-two eggs from the three women, and that they fertilized twenty-one of the eggs. Three of the eggs, one from each woman, developed into embryos and were transferred into the women's uteruses. None of the pregnancies lasted, but Cohen said it is far too soon to be discouraged. "This by no means means the method doesn't work," he told me.

Capron said that the strangest proposal he heard was from a doctor who dreamed of cloning to create a source of eggs for women whose ovaries had failed. The plan was to add the genes from one of the woman's cells to an enucleated egg from a donor. The doctor would allow the fetus to develop, then abort it and remove its ovaries. He would harvest eggs from the fetal ovaries, which would, of course, be genetically identical to the woman's eggs, if she had been able to make them. Then he would fertilize the fetal eggs in the lab, allowing the woman to have her own genetic children.

Although many aspects of this proposal are still at the stage of animal research, at best, the doctor argued, such treatment should not be banned, Capron reported. He said that, after all, there would be no cloned human beings resulting from the treatment. Of course, abortion opponents would object to the idea of a making a cloned fetus, only to abort it, martyring a fetus to allow a woman to have babies.

While the ideas may seem risky and futuristic, some infertility specialists passionately argued that those who pass judgment on the research may not understand what their patients are going through.

"None of these people have sat down and talked to my patients," said James Grifo, the director of the division of reproductive endocrinology at New York University Medical Center. "None of them have seen the misery my patients are living through."

"Who should make the ultimate decision?" Grifo asked. "Shouldn't patients be allowed to make these decisions for themselves?" Those who pass judgment on doctors who want to clone "leave out the interests of my patients," he said.

Most of these infertility experts, asked if humans will one day be cloned, replied, "Of course."

"There are advantages to cloning-related technology," Cohen told me. "There will be more and more applications," as time goes on, he predicted. "It is up to politicians and lawmakers to make up their minds whether asexual reproduction is acceptable. "In my opinion, it is all nonsense whether it is sexual or asexual," he added. "The whole argument is sort of silly—so what? People will agree with me in fifty years."

The most startling suggestion I heard was from Steen Willadsen. Willadsen now works part-time in Cohen's infertility center, where he does research with mouse eggs and eggs from infertile women that would otherwise have been discarded. He also works part-time at an infertility clinic near his home in Florida, where he injects sperm from infertile men directly into eggs in order to fertilize the eggs. These are sperm that ordinarily would rarely, if ever, have been able to penetrate an egg. Willadsen told me that he thinks humans may have already been cloned—accidentally.

Willadsen noted that the sperm-injection method is not used just for mature sperm, removed from semen. Infertility specialists can now fertilize eggs by injecting them with immature sperm taken directly from a man's testicles. The immature sperm have no tails and, in fact, look just like other cells that may be mixed in with them. They have just one set of chromosomes, like mature sperm or mature eggs. But, Willadsen said, other cells in the same sample can have two sets of chromosomes, as do most other cells in the human body. If the embryologist inadvertently injects a non-sperm cell into an egg, and if the egg, as sometimes happens, spontaneously shed its own nucleus, the procedure would be transformed into a cloning of the father.

"The likelihood of a baby resulting is small, but in the light of the Dolly experience, not nil," Willadsen said. Not many babies have been born from the direct injection of eggs with immature sperm, and there is a very low probability that the sperm injection will result in a clone. But there is a law of statistics that says that even the most unlikely event will eventually occur if you wait long enough. It is unlikely that a man has already been cloned from the sperm-injection method, but as more and more babies are created this way, sooner or later, it will result in a clone, Willadsen asserted.

Sauer feels that the public is not going to easily accept cloning, with its connotations of science fiction horrors. "I happen to believe that cloning is a politically dirty word," Sauer said. "I can't believe that any politician will stand up and say, 'Let's start cloning.'" But, he added, "among those who are capable of cloning, I don't think there's that sort of outrage."

In the end, it may come down to a matter of semantics, Willadsen told me. Yes, he is pretty sure that humans will be intentionally cloned one day. But, he said, "it probably will not be called cloning."

N O T E S

Notes reflect published sources only. Information derived from personal interviews is attributed to the source in the text.

CHAPTER 1: A CLONE IS BORN

Page

3   *"But could one really visualize the cell forming a finger . . .":* Willard Gaylin, "The Frankenstein Myth Becomes a Reality: We Have the Awful Knowledge to Make Exact Copies of Human Beings," *The New York Times Magazine,* March 5, 1972.

6   *Is it possible, as molecular biologist Gunther Stendt once suggested . . . :* Gunther Stendt, "Molecular Biology and Metaphysics," *Nature,* April 26, 1974, p. 781.

7   *"far less a surrender to the mystery of the genetic lottery,":* Gilbert Meilaender, testimony before the National Bioethics Advisory Commission, March 13, 1997.

7   *Elliott Dorff, a rabbi at the University of Judaism . . . :* Elliott Dorff, testimony before the National Bioethics Advisory Commission, March 14, 1997.

12 *"We can now take 5 cc of a woman's blood . . .":* Moshe Tendler, testimony before the National Bioethics Advisory Commission, March 14, 1997.

15 *He quoted the theologian Paul Ramsey:* Leon Kass, testimony before the National Bioethics Advisory Commission, March 13, 1997.

16 *The priest, Albert Moraczewski, of the National Conference on Catholic Bishops . . . :* Albert Moraczewski, testimony before the National Bioethics Advisory Commission, March 13, 1997.

17 *The next day, Moshe Tendler, an Orthodox Jewish rabbi . . . :* Moshe Tendler, testimony before the National Bioethics Advisory Commission, March 14, 1997.

18 *Nancy Duff, a theologian at the Princeton Theological Seminary . . . :* Nancy Duff, testimony before the National Bioethics Advisory Commission, March 14, 1997.

18 *John Robertson, a law professor at the University of Texas . . . :* John Robertson, testimony before the National Bioethics Advisory Commission, March 14, 1997.

20 *Ruth Macklin, an ethicist at Albert Einstein College of Medicine . . . :* Ruth Macklin, testimony before the National Bioethics Advisory Commission, March 14, 1997.

20 *The clashing viewpoints, said Ezekiel J. Emanuel . . . :* Ezekiel J. Emanuel, testimony before the National Bioethics Advisory Commission, March 14, 1997.

CHAPTER 2: BREAKING THE NEWS

23 *Wilmut is an Englishman . . . :* Michael Specter with Gina Kolata, "A New Creation: The Path to Cloning," *The New York Times,* March 3, 1997, p. A1.

23 *The Elizabethan poet William Drummond . . . :* "The Silence of the Lamb" in Talk of the Town, *The New Yorker,* March 18, 1997, p. 48.

24 *"I am not a fool," he said:* Michael Specter with Gina Kolata, op. cit.

33 *The article said that the only questions for* Time *and* Newsweek *. . . :* Kurt Andersen, "The Outsider," *The New Yorker,* March 31, 1997, p. 45–46.

34 *A week after the media swarm . . . :* "The Silence of the Lamb," op. cit.

CHAPTER 3: NATURAL PHILOSOPHIES

42 *The Roman orator Seneca . . . :* Quoted in Leon Eisenberg, *The Journal of Medicine and Philosophy,* vol. 1, no. 4, 1976, p. 320.

43 *After all, he wrote, "some of the parts are clearly visible . . .":* Quoted in "Aristotle: The Generation of Animals," from *The Philosophy of Biology,* ed. Michael Ruse (New York: Macmillan Publishers, 1989), pp. 28–31. Original passages are from Jonathan Barnes, ed., *The Complete Works of Aristotle: The Revised Oxford Translation,* Bollingen Series 71. (Jowett Copyright Trustees, 1984), pp. 1138–1141.

43 *the first scientists who looked at sperm . . . :* Leon Eisenberg, op. cit.

45 *"To biology students of my generation . . .":* Viktor Hamburger, *The Heritage of Experimental Embryology: Hans Spemann and the Organizer* (New York: Oxford University Press, 1988), p. vii.

47 *"We spent the equivalent of only a few dollars . . .":* Ibid., p. 21.

47 *Spemann wrote that making instruments . . . :* Ibid.

49 *"Again and again," Spemann wrote . . . :* Hans Spemann, *Embryonic Development and Induction* (New Haven, Conn.: Yale University Press, 1938), pp. 371–372.

52 *Criticizing Lamarck's theories . . . :* William James, *Principles of Psychology,* reprinted by Encyclopaedia Britannica, Inc., Great Books of the Western World, Robert Maynard Hutchins, editor-in-chief, vol. 53 (Chicago: University of Chicago Press, 1952), p. 896.

52 *"The fact that few Nobel Prizes have been conferred . . .":* Robert Gilmore McKinnell, *Cloning. A Biologist Reports* (University of Minnesota Press, 1979), p. 5.

54 *"The work done by Roux and others . . .":* Sigmund Freud, *General Introduction to Psycho-Analysis,* p. 595.

56 *Hamburger wrote that his friend . . . :* Viktor Hamburger, op. cit., p. 15.

57 *"The most accurate explanation for Roux's results . . .":* Robert Gilmore McKinnell, *Cloning of Frogs, Mice, and Other Animals* (University of Minnesota Press, 1979), p. 26.

57 *"I found here a theory of heredity . . .":* Viktor Hamburger, op. cit., p. 9.

58 *Spemann wrote of his experiences . . . :* Viktor Hamburger, op. cit., p. 11.

61 *"The first half of this experiment . . .":* Hans Spemann, op. cit., pp. 210–211.

63 *One reviewer commented . . . :* Marie A. Di Berardino, *Genomic Potential of Differentiated Cells* (New York: Columbia University Press, 1997), p. 35.

64 *"We gently patted Tom on the shoulder . . .":* Ibid., p. 45.

64 *'Twas sometime before Christmas in the year '51 . . . :* Ibid., pp. 46–47.

66 *"The cloning of adults (human and nonhuman) . . . ":* Robert Gilmore McKinnell, op. cit., p. 55.

67 *In 1962, in a famous experiment . . . :* John B. Gurdon, "Adult Frogs Derived from the Nuclei of Single Somatic Cells," *Developmental Biology,* vol. 4, 1962, pp. 256–273.

CHAPTER 4: IMAGINING CLONES

72 *If we cloned people with "attested ability" . . . :* J.B.S. Haldane, "Biological Possibilities for the Human Species in the Next Ten Thousand Years," *Man and His Future, a CIBA Foundation Volume,* ed. Gordon Wolstenholme (J&A Churchill Ltd., 1963), p. 352.

72 *"For exceptional people commonly have unhappy childhoods . . .":* Ibid.

73 *"It is an interesting exercise in social science fiction . . .":* Joshua Lederberg, *The Washington Post,* September 30, 1967.

76 *"Just as every child must have the right to full educational opportunity . . .":* Bentley Glass, *Science,* vol. 171, no. 3966, January 8, 1971, p. 28.

76 *In a paper published in 1968 . . . :* Linus Pauling, *UCLA Law Review,* vol. 15:267, 1968, p. 269.

78 *And it sowed such a profound distrust . . . :* Lynda Richardson, "Experiment Leaves Legacy of Distrust of New AIDS Drugs," *The New York Times,* March 21, 1997, p. A1.

81 *"It's a foreboding I have . . .":* Carl Sagan, *Skeptical Inquirer,* vol. 19, no. 1, January–February 1995, pp. 24–30.

81 *"We went to the Moon and all we got out of it was Teflon pans . . .":* Seth Schiesel, "Once Visionary, Disney Calls Future a Thing of the Past," *The New York Times,* February 23, 1997, p. 24.

81 *Futurist Alvin Toffler . . . :* Ibid.

83 *One writer asked . . . :* "The Mail," *Atlantic Monthly,* July 1971, p. 26.

85 *"The study of Nature . . .":* H. G. Wells, *Seven Science Fiction Novels* (Dover Press, 1934), p. 134.

88 *"I do not believe that nuclear transplantation . . .":* Robert Gilmore McKinnell, *Cloning. A Biologist Reports* (University of Minnesota Press, 1979), p. 102.

89 *"It is almost no comfort to know that one's cloned, identical surrogate lives on . . .":* Lewis Thomas, *New England Journal of Medicine,* December 12, 1974, p. 1296.

89 *"Cloning from adult mammals may therefore remain science fiction.":* Bernard Davis, *The Genetic Revolution: Scientific Prospects and Public Perception,* ed. Bernard Davis (Baltimore, Md.: Johns Hopkins University Press, 1991), pp. 251–252.

90 *"It is unfortunate that Dr. Lederberg is either unaware or unwilling . . .":* Leon Kass, Letters to the Editor, *The Washington Post,* November 3, 1967, p. A20.

90 *"human procreation has already been replaced . . .":* Paul Ramsey, *Journal of the American Medical Association,* vol. 220, no. 11, June 12, 1972, p. 1980.

91 *"Given the intricacies of the human mind . . .":* Editorial, ibid., p. 1357.

91 *One of the final papers of this flurry of interest . . . :* Gunther Stendt, "Molecular Biology and Metaphysics, *Nature,* April 26, 1974, p. 780.

CHAPTER 5: THE SULLYING OF SCIENCE

96 *Rorvik claimed that one day in September of 1973 . . . :* David M. Rorvik, *In His Image: The Cloning of a Man* (New York: J. B. Lippincott, 1978), p. 23.

96 *Rorvik agonized over the proposal . . . :* Ibid., p. 25.

96 *"I knew that, if my part in this came to light . . .":* Ibid., p. 28.

97 *But, he wrote, he managed to win . . . :* Ibid., p. 44.

98 *But he was uncomfortable, Rorvik reported . . . :* Ibid., p. 80.

99 *Roberto, Rorvik wrote, "would go through the factories . . .":* Ibid., p. 124.

99 *She had the baby "in a small hospital." . . . :* Ibid., p. 204.

100 *His publisher issued a brief statement . . . :* Lee Lescaze, "A Baby Book of a

Different Nature: Writer's Claim That Scientist Cloned a Child Rekindles Debate," *The Washington Post,* March 8, 1978, p. A3.

100 *Jonathan Segal, then a senior editor at Simon & Schuster . . . :* Ibid.

100 *For example,* Science *reported, some said the book could not be true . . . :* Barbara J. Culliton, "Scientists Dispute Book's Claim That Human Clone Has Been Born," *Science,* March 24, 1978, p. 1316.

100 Newsweek *reported in its March 20 issue . . . :* Peter Gwynne, "All About Clones," *Newsweek,* March 20, 1978, p. 68.

101 *"If Hitler were cloned . . .":* Quoted in ibid.

101 *Beckwith said, "Even if this is a hoax . . .":* Quoted in Barbara J. Culliton, "Cloning Caper Makes It to the Halls of Congress," *Science,* March 24, 1978, p. 1316.

102 *A spokesman for the health subcommittee . . . :* Quoted in Barbara J. Culliton, "Scientists Dispute Book's Claim That Human Clone Has Been Born," op. cit.

102 *"Put yourself in this position . . .":* Quoted in Lee Lescaze, op. cit.

102 *"Every time someone cries hoax, he's delighted," Rorvik said.:* Ibid.

102 *Max Lerner, in a column in* The Washington Post *. . . :* Max Lerner, *The Washington Post,* March 31, 1978.

103 *"It is nevertheless mildly amusing . . .":* Beatrice Mintz, testimony before the Subcommittee on Health and the Environment of the Committee on Interstate and Foreign Commerce, House of Representatives, Ninety-fifth Congress, second session, May 31, 1978, p. 7.

103 *He said that since cloning experiments . . . :* Thomas Briggs, testimony before the Subcommittee on Health and the Environment, p. 11.

103 *Clement Markert of Yale . . . :* Clement Markert, testimony before the Subcommittee on Health and the Environment, p. 18.

104 *"I enjoyed it particularly because I knew it to be nonsense . . .":* Andre Helligers, testimony before the Subcommittee on Health and the Environment, p. 87.

104 *"I entitle this book not* In His Image *. . .":* Andre Helligers, testimony before the Subcommittee on Health and the Environment, p. 89.

104 *James Watson, who just a few years earlier had written an article . . . :* Interview by C. P. Anderson, "In His Own Words: Nobel Laureate James Watson Calls Report of Cloning People 'Science Fiction Silliness,' " *People,* April 17, 1978, pp. 93–95.

104 *" 'Bunk,' say numerous scientists . . .":* Herschel Johnson, "Cloning: Can Science Make Copies of You?," *Ebony,* July 1978, p. 96.

107 *Three senior scientists at the Massachusetts Institute of Technology . . . :* Joshua Lederberg, "Spreading Research Strikes Score on Wrong Target," *The Washington Post,* March 1, 1969.

108 *The first Earth Day was proclaimed in 1970. . . . :* James Watson and John Tooze, *The DNA Story: A Documentary History of DNA Cloning* (New York: W. H. Freeman and Co., 1981), p. viii.

108 *"Our first reaction was pure joy," Watson wrote:* Ibid., p. vii.

109 *"Some of the more panicky reactions to genetic engineering . . .":* Joshua Lederberg, *The Washington Post,* November 4, 1969.

110 *Others, like Pollack, feared that scientists . . . :* William Bennett and Joel Gurin, "Science That Frightens Scientists: The Great Debate over DNA," *Atlantic Monthly,* February 1977, p. 45.

110 *"He was absolutely dumbfounded, as far as I could see," . . . :* Ibid., p. 46.

110 *Wallace Rowe, a molecular biologist . . . :* Quoted in James D. Watson and John Tooze, op. cit., p. 2.

110 *"Have we the right to counteract . . .":* Irwin Chargaff, Letters to the Editor, *Science,* June 4, 1976, p. 938.

111 *"Are we really that much farther along on the path . . .":* Philip Siekevitz, Letters to the Editor, *Science,* October 15, 1976, p. 257.

111 *"I claim that the exploitation of recombinant DNA . . .":* Freeman Dyson, Letters to the Editor, *Science,* July 2, 1976, p. 6.

112 *He attributed part of the fear of molecular biology . . . :* Horace Freeland Judson, "Fearful of Science," *Harper's,* vol. 250. no. 1498, March 1975, p. 36.

112 *Once again, the feuding scientists emerged:* Nicholas Wade, "Recombinant DNA: New York State Ponders Action to Control Research," *Science,* November 12, 1976, p. 705.

113 He wrote, "The pure research potential . . .": Lewis Thomas, "Hubris in Science?," *Science*, 1978, p. 1459.

113 "Ever since we achieved a breakthrough . . .": Interview by C. P. Anderson, op. cit., p. 96.

113 Genentech, a company founded by Stanford scientists . . . : James D. Watson and John Tooze, op. cit., p. ix.

116 Bromhall wanted a court order . . . : Jim Quinn, "$7 Million Libel Suit Filed over Clone Book," *The Washington Post*, July 11, 1978, p. A3.

119 "It was a big stiff," said Richard Clark . . . : Quoted in Ebet Roberts, "Rorvik: Still Cloning Away," *Newsweek*, January 14, 1980, p. 17.

CHAPTER 6: THREE CLONED MICE

123 He managed to suck the nuclei out of rabbit eggs . . . : J. Derek Bromhall, "Nuclear Transplantation in the Rabbit Egg," *Nature*, vol. 258, 1975, pp. 719–721.

124 The Jackson Laboratory is the world's only nonprofit research institution . . . : Lee M. Silver, *Mouse Genetics* (New York: Oxford University Press, 1995), pp. 9–10.

125 Illmensee had completed an experiment . . . : Karl Illmensee, "Nuclear and Cytoplasmic Transplantation in Drosophila," *Insect Development*, ed. P. A. Lawrence (Blackwell Scientific, 1976), pp. 76–96.

127 Illmensee and Hoppe said that they had created five fatherless mice . . . : Peter C. Hoppe and Karl Illmensee, "Microsurgically Produced Homozygous-Diploid Uniparental Mice," *Proceedings of the National Academy of Sciences*, vol. 74, no. 12, December 1977, pp. 5657–5661.

127 Others, including leading scientists like Clement Markert . . . : Clement L. Markert and R. M. Peters, *Journal of Experimental Zoology*, vol. 201, 1977, pp. 295–302.

128 U.S. News & World Report announced . . . : "Scientific Feat: Test-Tube Mice," *U.S. News & World Report*, January 19, 1981, p. 7.

128 Newsweek reporter Sharon Begley asked . . . : Sharon Begley, "The Three Cloned Mice," *Newsweek*, January 16, 1981, p. 65.

128 Charles Krauthammer, writing in The New Republic . . . : Charles Krauthammer, "Tales from the Hatchery," *The New Republic*, February 14, 1981, pp. 12–13.

137   *When he finished, Bürki stood up . . . :* "Report to the International Commission of Inquiry into the Scientific Activities of Prof. Karl Illmensee," Geneva, January 30, 1984, p. 1.

137   *Illmensee "presented females that had not been mated the previous night" . . . :* Ibid.

138   *"Therefore, in later discussions with Prof. Illmensee . . .":* Ibid., p. 3.

139   *"Dr. Illmensee clearly recognized . . .":* Ibid.

139   *He categorically rejected the charges against him . . . :* "Exclusive Interview with Karl Illmensee," *Bild der Wissenshaft,* August 1983, pp. 89–96.

141   *The professors who were with Illmensee when he signed . . . :* "Report to International Commission of Inquiry," p. 19.

141   *He had only meant to convey . . . :* Ibid., p. 20.

141   *The committee wrote that Illmensee had made . . . :* Ibid., pp. 16–17.

142   *But he lost his research grants . . . :* Stephen Budiansky, "NIH Withdraws Research Grant," *Nature,* June 28, 1984, p. 734.

147   *"I am fully convinced of his personal and scientific integrity . . .":* Eckhard Lieb, "Unfair on Illmensee," *Nature,* June 21, 1984, vol. 309, p. 664.

147   *David Yaffee, a cell biologist at the Weizmann Institute of Science . . . :* David Yaffee, "Prejudiced Reporting?," *Nature,* vol. 305, September 15, 1983, p. 176.

147   *Hans J. Becker, a biology professor . . . :* Letter supplied by Davor Solter.

154   *He faxed me a copy of a letter to the University of Geneva . . . :* Letter of May 20, 1991, to Le Recteur, Université de Geneva, from R. L. Gardner and A. McLaren, Imperial Cancer Research Fund and Oxford University.

CHAPTER 7: BREAKING THE LAWS OF NATURE

167   *In March 1986, Steen Willadsen in England announced . . . :* Steen M. Willadsen, "Nuclear Transplantation in Sheep Embryos," *Nature,* vol. 320, March 6, 1986, pp. 63–65.

167   *Prather, Eyestone, and a handful of other young doctoral students . . . :* R. S. Prather et al., "Nuclear Transplantation in the Bovine Embryo: Assessment of Door Nuclei and Recipient Oocyte," *Biology of Reproduction,* vol. 37, November 1987, pp. 859–866.

175 *Like Ian Wilmut . . . :* Steen M. Willadsen et al., "Deep Freezing of Sheep Embryos," *Journal of Reproduction and Fertility,* vol. 46, 1976, pp. 151–154.

176 *One day, he hit on a solution. . . . :* S. M. Willadsen, "The Developmental Capacity of Blastomeres from 4- and 8-Cell Sheep Embryos," *Journal of Embryology and Experimental Morphology,* vol. 65, 1981, pp. 165–172.

177 *He asked himself how many times . . . :* S. M. Willadsen and C. Polge, "Attempts to Produce Monzygotic Quadruplets in Cattle by Blastomere Separation," *Veterinary Record,* vol. 114, 1984, pp. 240–243.

180 *The creature was strange indeed . . . :* Carole B. Fehilly, Steen M. Willadsen, and Elizabeth M. Tucker, "Interspecific Chimaerism Between Sheep and Goat," *Nature,* March 12, 1984, pp. 634–636.

CHAPTER 8: THE ROAD TO DOLLY

186 *As evidence of what lay ahead . . . :* Keith Schneider, "Better Farm Animals Duplicated by Cloning," *The New York Times,* February 17, 1988, p. A1.

208 *Campbell and Wilmut wrote a paper for* Nature *. . . :* K.H.S. Campbell et al., "Sheep Cloned by Nuclear Transfer from a Cultured Cell Line," *Nature,* vol. 380, March 7, 1996, pp. 64–66.

208 *This time, Solter ended his editorial very differently . . . :* Davor Solter, "Lambing by Nuclear Transfer," *Nature,* vol. 380, March 7, 1996, p. 25.

CHAPTER 9: TAKEN BY SURPRISE

217 *For cloning, they would use three types of cells . . . :* I. Wilmut, et al., "Viable Offspring Derived from Fetal and Adult Mammalian Cells," *Nature,* February 27, 1997, p. 811.

218 *In keeping with Wilmut and Campbell's suspicions . . . :* Ibid.

CHAPTER 10: THE PATH AHEAD

228 *He stood on a stage at a public forum on cloning . . . :* Wilmut spoke at a public forum on cloning held by the American Association for the Advancement of Science on June 26, 1997.

229 *"The strong public reaction suggests a strong argument . . .":* Emanuel spoke at a public forum on cloning held by the American Association for the Advancement of Science on June 26, 1997.

230 *Its members, an assortment of eminent scientists . . . :* "Declaration in Defense of Cloning and the Integrity of Scientific Research," *Free Inquiry,* Summer 1997, pp. 11–12.

230 *"The moral issues raised by cloning . . .":* Ibid.

230 *Richard Dawkins contributed an accompanying essay . . . :* Richard Dawkins, "Thinking Clearly About Clones," op. cit., pp. 13–14.

231 *"The potential benefits of cloning may be so immense . . .":* "Declaration in Defense of Cloning and the Integrity of Scientific Research," op. cit., p. 12.

235 *In June 1991, Jay Feinberg . . . :* Gina Kolata, "Despite Scandals, Research Programs Thrive, *The New York Times,* May 25, 1994, p. A16.

236 *Ultimately, Orkin said, scientists might be able to bypass . . . :* Stuart Orkin, testimony before the National Bioethics Advisory Commission, March 14, 1997.

241 *Elizabeth H. Blackburn, a scientist at the University of California . . . :* Gina Kolata, "Scientists Rethinking the Role of Chromosomal 'Leader Tape,'" *The New York Times,* February 25, 1997, p. C3.

242 *But, at least in the private world of infertility clinics . . . :* Gina Kolata, "For Some Infertility Experts, Human Cloning Is a Dream," *The New York Times,* June 7, 1997, p. A8.

INDEX

embryos of, 159–160, 163–168,
185–190, 191, 192,
193–194, 199, 215–216
mad cow disease in, 26–27, 226
prize dairy, 10, 30, 40
Cavalieri, Liebe, 101, 112
Cell, 128, 146, 148, 173
cell cycle, 200–201, 203–207,
218, 220
cells, 32, 35, 50, 151
adult, 25–26, 29–30, 40, 71, 123,
147, 168, 198–199, 203,
212–213, 216–218,
225–226, 233, 239–242,
245
carrot root, 86–87
differentiation of, 24–25, 29–30,
44–45, 50–61, 62, 65, 66,
67, 73, 86, 186–187,
200–201, 204–208, 212
division of, 200–201, 204–205,
240
embryo, *see* embryo cells
embryonic stem (ES), 197–199,
203, 205, 206, 212
fetal, 198–199, 203, 205, 212,
232
grown in laboratory, 43–44,
86–87, 130–131, 163, 198
meiosis of, 68
separation of, 86, 187, 203
specialized, 63, 66, 67–68, 69,
97, 103, 133, 182,
186–187, 198, 204
unidirectional development of,
50–51, 73, 89
see also cytoplasm; eggs; sperm
Central Intelligence Agency, 101
Chambon, Pierre, 143
Chargaff, Erwin, 110–111
Charo, R. Alta, 37–38

chimeras, 179–180, 184, 195, 197
chromosome diminution, 53
chromosomes, 24, 68
abnormalities of, 238–239, 243
telomeres of, 240–241
of worm (*Ascaris megalocephala*),
52–53
Clark, Richard, 119
Clinton, Bill, 35, 228–229, 230,
233
"clone," as term, 71–72
Clone Rights United Front, 35
cloning, 1–21
accidental, 247–248
in advertisements, 34
of the dead, 3, 32, 40
difficulty of, 24–25, 29, 40,
65–67, 71, 73, 88, 89, 93
environmental influences and,
37–38, 50, 83, 87, 88
of exceptional people, 72, 83,
91–92, 114, 128
genetic parent in, 27–28, 32,
207, 216–217
jokes about, 34
methodology of, 3, 25–26,
27–28, 30, 61, 64
in movies, 189–190
multiple, 3–4, 91–92, 159–160,
176–178, 179–180,
185–186, 191, 242–243
of oneself, 3–4, 5–6, 8–9, 16, 19,
20, 35, 85, 87, 89,
92–105, 116–119, 230–232
potential benefits of, 8–10, 17,
30–31, 39, 71–74,
229–237
as production vs. procreation,
6–7, 19, 90–91
in science fiction, 95, 113–115,
190, 248